Ageing with Smartphones in Japan

Ageing with Smartphones in Japan

Care in a visual digital age

Laura Haapio-Kirk

ⒶUCLPRESS

First published in 2024 by
UCL Press
University College London
Gower Street
London WC1E 6BT

Available to download free: www.uclpress.co.uk

ISBN: 978-1-78735-578-1 (Hbk)
ISBN: 978-1-78735-577-4 (Pbk)
ISBN: 978-1-78735-576-7 (PDF)
ISBN: 978-1-78735-579-8 (epub)
DOI: https://doi.org/10.14324/111.9781787355767

For Suvi

Contents

List of figures

Preface

Names of research participants are written with the family name first followed by the given name, as is the custom in Japan. The exception to this is in cases where an individual has requested to be identified by their given name first, or by their given name only. Pseudonyms are used throughout, except in cases where participants specifically requested the use of their real names. The common honorific '-san' is used in all cases. Romanisation of Japanese words is used throughout, following the Hepburn system. Macrons (indicating long vowels) have been avoided in those cases where the romanised spelling ignores them, for example for principal cities (Kyoto rather than Kyōto, Osaka rather than Ōsaka). The exchange rate during my fieldwork between 2018–19 was roughly 140 ¥ (JPY) to £1 (GBP).

The illustrations that appear at the beginning of each chapter are intended to offer a moment of pause and to inspire connections between the book's overall themes. Deliberately without captions, they invite the reader's own interpretation, mimicking ethnographic fieldwork in which meaning is sought from experiences and observations, often after dwelling with them for some time. As such, the images are not included in the List of figures. All the chapter frontispiece drawings are by Laura Haapio-Kirk.

Series foreword

This book series is based on a project called ASSA – The Anthropology of Smartphones and Smart Ageing. It was primarily funded by the European Research Council (ERC) and located at the Department of Anthropology, UCL. The project had three main goals. The first was to study ageing. Our premise was that most studies of ageing focus on those defined by age, that is youth and the elderly. This project would focus upon people who did not regard themselves as either young or elderly. We anticipated that their sense of ageing would also be impacted by the recent spread of smartphone use. Smartphones were thereby transformed from a youth technology to a device used by anyone. This also meant that, for the first time, we could make a general assessment of the use and consequences of smartphones as a global technology, beyond those connotations of youth. The third goal was more practical. We wanted to consider how the smartphone has impacted upon the health of people in this age group and whether we could contribute to this field. More specifically, this would be the arena of mHealth, that is, smartphone apps designed for health purposes.

The project consists of 11 researchers working in 10 fieldsites across nine countries, as follows: Al-Quds (East Jerusalem) studied by Laila Abed Rabho and Maya de Vries; Bento, in São Paulo, Brazil studied by Marília Duque; Cuan in Ireland studied by Daniel Miller; Kampala, Uganda studied by Charlotte Hawkins; rural Kōchi and urban Kyoto in Japan studied by Laura Haapio-Kirk; NoLo in Milan, Italy studied by Shireen Walton; Santiago in Chile studied by Alfonso Otaegui; Shanghai in China studied by Xinyuan Wang; Thornhill in Ireland studied by Pauline Garvey; and Yaoundé in Cameroon studied by Patrick Awondo. Several of the fieldsite names are pseudonyms.

Most of the researchers are funded by the European Research Council. The exceptions are Alfonso Otaegui, who is funded by the Pontificia Universidad Católica de Chile, and Marília Duque, Laila Abed Rabho and Maya de Vries, who are mainly self-funded. Pauline Garvey is based at Maynooth University. The research was simultaneous except for the research in Al-Quds, which has been extended since the researchers are also working as they research.

The project has published a comparative book about the use and consequences of smartphones called *The Global Smartphone*. In addition, we intend to publish an edited collection presenting our work in the area of mHealth. There are also nine monographs representing our ethnographic research, the two fieldsites in Ireland combined in a single volume. These ethnographic monographs mostly have the same chapter headings. This will enable readers to consider our work comparatively. The project has been highly collaborative and comparative from the beginning. We have been blogging since its inception at https://blogs.ucl.ac.uk/assa/. Further information about the project may be found on our project's main website, at https://www.ucl.ac.uk/anthropology/assa/. The core of this website is translated into the languages of our fieldsites and we hope that the comparative book and the monographs will also appear in translation. As far as possible, all our work is available without cost, under a Creative Commons licence.

Acknowledgements

First and foremost, I express my deepest gratitude to all the people who shared their lives with me during the fieldwork that produced this book. I am grateful for their friendship and for trusting me with their stories. I hope that I have done them justice. There are many anonymous people to whom I owe a heartfelt thank you – their contribution cannot be underestimated. I thank the staff at Tosa-chō Town Hall, in particular Toriyama Yuriko and Ishikawa Takashi for their help and generosity. I also thank Yamakubi Naoko at the Tosa-chō Social Welfare Council for her care and assistance. Thank you to Ito Megumi who welcomed me to Tosa-chō and whose beautiful painting features in Chapter 8.

I owe my most sincere thanks to Sasaki Lise for her research assistance, friendship and constant support throughout this project and beyond. I could not have produced this research without her. Her expertise and tireless enthusiasm continue to inspire me. Thank you to the Sasaki family for welcoming me into their home, and especially to Hiromi-san for contributing her wonderful drawing, featured in Chapter 5.

During fieldwork I benefited from being affiliated as a visiting researcher at Osaka University, under the mentorship of Beverly Yamamoto. I would like to thank all of the people I met there who provided intellectual engagement with my work. In particular, I am grateful to Kimura Yumi who invited me to join the Field Medicine team of researchers at the rural health check in Tosa-chō, led by Sakamoto Ryota. Thank you to Andrea De Antoni at Kyoto University and to members of the Emosense group who gave valuable feedback on material I presented during fieldwork. I am grateful to the University of Jyväskylä for supporting a three-month fellowship during the writing-up of this research, during which I received helpful comments from Sakari Taipale, Riitta Hänninen and other members of the CoE AgeCare. Thank you to Christ Church, Oxford for supporting me during the writing of this book through a Junior Research Fellowship.

I have been extremely fortunate to be a member of the Anthropology of Smartphones and Smart Ageing project. I would like to thank the ERC for funding the research and all of the team members for regularly reading my work, as well as for providing constant inspiration, support and friendship. My thanks to Laila Abed Rabho, Patrick Awondo, Maya de Vries, Marília Duque, Pauline Garvey, Charlotte Hawkins, Daniel Miller, Alfonso Otaegui, Shireen Walton, Xinyuan Wang and Georgiana Murariu. I am also grateful to the members of the DDD group in Oxford who generously read and commented on my work, in particular to Charlotte Linton for stimulating discussions about rural Japan. I wish to thank Jolynna Sinanan who generously read my PhD thesis when it was still in an embryonic stage. She provided me with much-needed encouragement at a critical moment and a clear path to completion. Thanks also to the AnthroCo-write group for all their support.

I am profoundly grateful to my PhD supervisors Danny Miller and Inge Daniels for their unwavering support and ongoing guidance. Thank you to Jamie Coates and Liz Hallam for the enjoyable and insightful discussion as examiners for my PhD viva, which subsequently helped me to write this book. Thank you to Emily Haworth-Booth for her mentorship regarding graphic storytelling.

I am immensely grateful to the two anonymous reviewers who provided thoughtful, generous and constructive feedback on the book in typescript form. However, I take full responsibility for any inaccuracies or errors that remain. Thank you to the editorial team at UCL Press for their patience and guidance, and to copyeditor Catherine Bradley for her work on the manuscript.

Finally, I would like to thank my family and friends for their support throughout this journey. I especially thank my mother, Joy Kirk, for her devoted care. Thank you to Andrew Cropper, who encourages me to dream big and makes me laugh every day. Finally, thank you to Suvi, who arrived just after I submitted my PhD, for being a constant source of delight and teaching me new dimensions of care as I went on to complete this book.

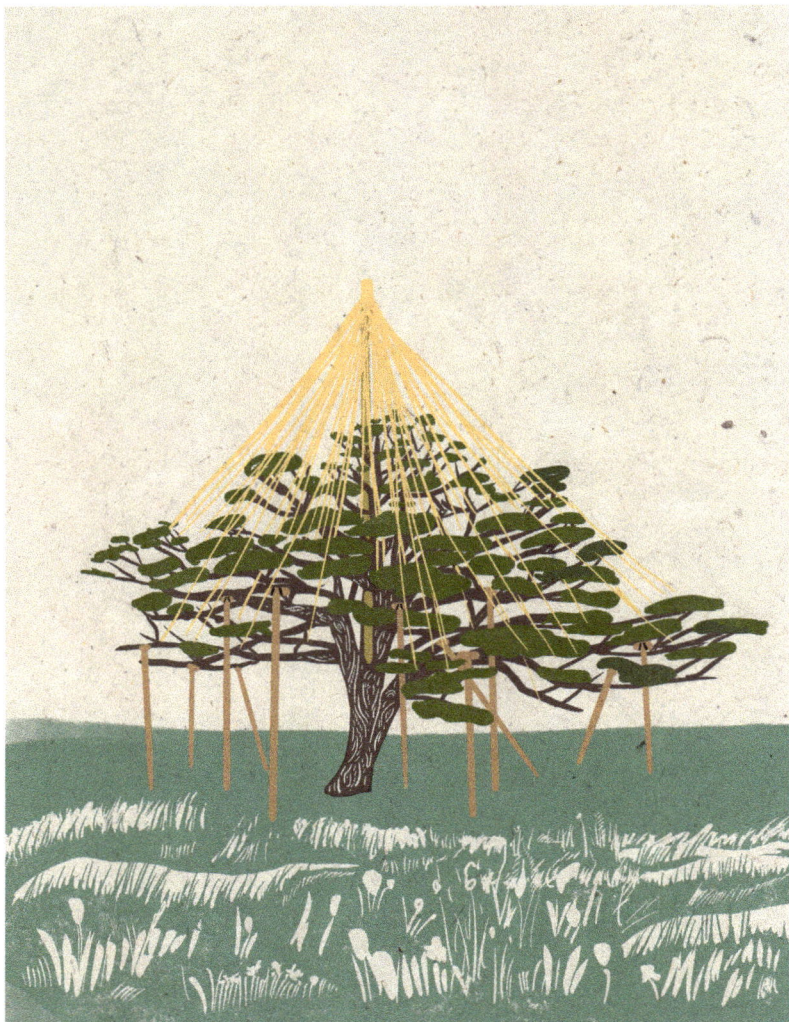

Image by Laura Haapio-Kirk.

1
Introduction

When walking around a garden in Japan, it is not unusual to see the twisting boughs of old trees propped up by wooden supports. It can be hard to see where the tree ends and the supports begin. Sometimes a particularly precarious branch may have several supports along its length. A great deal of care and effort goes into maintaining the longevity of old trees which are revered for their resilience in Shinto mythology, one of the core religions of Japan alongside Buddhism.[1] The Kenrokuen gardens in Kanazawa are famous for going the extra mile in this respect. Their pines are held aloft from below by a thick forest of sturdy wooden beams; in winter they receive extra support from above via bamboo poles and ropes tied in conical shapes above their foliage, known as *yukitsuri*, to protect delicate leaves and limbs from heavy snow. As I marvelled at this extraordinary feat of engineering and care, I was struck by how it paralleled what I was witnessing among the research participants for this book. Many of the older people I met were enmeshed in support networks, both from below (from family, friends and communities) and from above (from institutional and state sources of care). This book is about another kind of technology of care, emerging as central to the delivery of support from both directions: the smartphone.

In a country where people are among the oldest in the world, it is no surprise that how best to care for people in the latter part of their lives is a topic of national concern. This book, rather than focusing on the 'oldest old', primarily features people around the age of retirement. As older people begin to adopt digital technologies, in particular the smartphone, this book asks what it means to age with smartphones and explores how the device is implicated in forms of care. Rather than focusing on the intense kind of care that can be required towards the end of life, this book takes a broader approach, examining care as a particular orientation to being in the world. It explores how care manifests in everyday interactions between people and through material culture, from smartphones to housing, highlighting its significance in shaping individual experiences and collective wellbeing.

For older adults going digital brings a multitude of repercussions on their relationships, living arrangements and possibilities for what later life can be. Yet at the same time digitalisation can leave others behind, compounding issues of isolation and inaccessibility. Simultaneously the repercussions of ageing with smartphones manifest in distinctive digital cultures, including a profusion of visual communication. The near ubiquity of smartphones among people of all ages around the world means that older adults are now shaping smartphone cultures as much as young people, both through their usage and in the way that technology companies are adapting to the growing 'silver market', especially in Japan.

This book is based on the 16 months of anthropological ethnographic fieldwork I conducted primarily in urban Kyoto and rural Shikoku, Japan, between 2018 and 2019. The central themes of the book revolve around how people think about ageing 'well', from developing communities of support to finding purpose in their lives, and how such experiences are heavily gendered. I examine how experiences of ageing are impacted by the uptake of the smartphone and the development of digital repertoires, showing that smartphones and digital technologies are not only an urban phenomenon for young people but are also increasingly essential for rural-dwelling older adults. I argue that visual digital communication in the form of stickers and emojis,[2] rather than being just cute embellishments to messages, in fact play an important role in replicating a sense of physical proximity to others in a landscape of increasing distance. Depopulation, a trend associated with the ageing society in Japan, means that such interactions take on new significance as manifestations of care, something that I argue is central to the very meaning and purpose of life – or *ikigai* in Japanese.

A new era

The majority of people in this book are aged between 45 and 85 years old. They were born during the *Showa* era (1926–1989) and lived through the traumas of the Second World War and Japan's post-war economic rise, which brought about radical changes in lifestyles and the course of their lives.[3] The subsequent *Heisei* era (1989–2019) was dominated by economic decline and the rise of the 'super-ageing' society. In May 2019 Japan prepared itself to enter a new era: *Reiwa*.[4] Both trends, depopulation and the rise of digital technology, are of course set to continue into the *Reiwa* era. However, as this book demonstrates, it is especially the older people in Japan who are wondering what the new era will bring as their lives become increasingly digital.

The name *Reiwa*, meaning 'beautiful harmony', was taken from an old Japanese poem about plum blossoms. Prime Minister Abe explained that the name was chosen for the way it communicates values of tradition, care and an orientation towards the future that he hoped would define the new era:

> Culture is born through the beauty of people caring for one another... After a cold winter, spring comes. Like beautiful plum blossoms, the Japanese people, with hope for tomorrow in their hearts, can each make flowers bloom.[5]

On the day that the new era began, 1 May 2019, I was in rural Kōchi Prefecture, in the small town of Tosa-chō where I had been conducting ethnography for the past eight months. In retrospect, that day presents a snapshot of life at this moment of transition. The people I spent time with on that day illustrate the variety of ways in which people are living the values spoken about by Abe: 'caring for one another' with hopefulness, but at the same time revealing how their choices are shaped by wider social contexts beyond their control.

I began the day by visiting two sisters in their sixties. They had decided to build a modern and spacious house to live in together upon their return to their hometown from living and working in separate cities. Both were single, and it made sense for them to pool their resources and build their dream home with impressive views over the rice terraces that cascaded down the mountain. When filled with water, the terraces reflected magnificent sunsets, of which they proudly showed me photographs. Their mountain-top home required digital connectivity. Amazon deliveries made it possible to live in a remote spot and they were hopeful that they might have Airbnb guests to stay with them in their spare bedroom. The women's happiness in retirement was amplified by the camaraderie of being together and of defying tradition. It did not matter that they were without the support of a husband's salary since their rural hometown offered low-cost living that enabled them to avoid some of the financial challenges facing older single women, as will be discussed in Chapter 3.

I then visited a woman in her late eighties called Chieno-san[6] who lived in a large modern home next to her son. Her family home was next to their cattle shed in which they kept the local breed of cow – a distinctive, reddish-brown specimen developed in Kōchi Prefecture and called *akaushi*. Her late husband had dedicated his life to raising these prize-winning cows and she proudly displayed the trophies in her home, along with photos of her family with the cows which lined the ceiling of her living room. When I first met Chieno-san she shared how she had witnessed the atomic bomb blast at Nagasaki. She often told this story, indicating its importance to her:[7]

> I saw the atomic bomb in Nagasaki. I was working with a fisher-man. I saw it! The bomb!! There was a warning. I wasn't in the bun-ker, so I saw it. The mushroom cloud, the light. I saw it so clearly. I was 18. It was scary! When that happened, I came straight back to Kōchi. The train was so packed. It was human over human [*sic*]. There were injured soldiers, people like me, people who were going back home. I came home in that. I'm happy I'm here. That's why I don't make a fuss (*kuyo kuyo*). I feel that I have travelled around the world. I saw the people who were affected by the bomb. When I came back, I went past Hiroshima. There was still smoke every-where. The town was gone.

On Chieno-san's table was spread that morning's newspaper. She turned to the front page to check how to write the Chinese characters (*kanji*) for the new era in order to sign a consent form I had given to her. On the front page, below the announcement of the *Reiwa* era, were pho-tos of destroyed towns in the Tōhoku region, highlighting the continual struggle for people to rebuild their lives after the tsunami of 11 March 2011 obliterated entire communities. Recalling Chieno-san's descrip-tion of the devastation following the nuclear attacks on Japan, the pho-tos were a stark reminder of the uprootedness and precarity[8] that still faces many people after this most recent cataclysmic event. As will be described in Chapter 7, the triple disaster of earthquake, tsunami and nuclear meltdown that occurred in 2011 (events that have come to be referred to as '3.11') motivated some people to move away from cities to build a more self-sustainable rural life, which paradoxically requires forms of dependence on the local community. The events of the Second World War still loomed large in Chieno-san's mind. They have shaped her into the kind of person she is today who 'doesn't make a fuss', in her words, because she has witnessed the horrors confronting so many people at that time.

The continued effects of 3.11 in the minds of many people in this research were similarly profound. People in Tosa-chō wondered about how cut off they would be if or when another big disaster hit Japan; they felt that they would be low in terms of governmental priority. The smartphone emerged as an important safety net in times of disaster. One man in his forties had a shortcut permanently on his smartphone home screen that took him directly to a website informing him of water levels in Kōchi – said to fluctuate before a large earthquake. If he picked up on warning signs, he was ready to evacuate with his family before, as he predicted, transport and other infrastructures were cut off to this rural region. People in the Kansai region of Western Japan, where I also conducted research, spoke of the predicted *Nankai* megathrust earthquake, which they anticipated with a sense of certainty and dread.

After Chieno-san's visit I had been invited to attend a ceremony welcoming the new era at a Shinto shrine, which coincided with a celebration for the 840-year anniversary of the shrine's founding. I was grateful to be so warmly received as an invited guest to the ceremony and subsequent meal with the community. However, I noticed that there were no urban-to-rural migrants present, confirming to me the divisions existing in this rural town between locals and those who have migrated from cities (so-called 'I-turners'). The term is used in contrast to 'U-turners', who are urban migrants returning to their rural roots. The 'U-turners' face less resistance to integration with the local community.[9]

In the evening I visited two 'I-turner' families. Both had young children and were of a similar age but their lifestyles were very different, showing the diversity that exists among such migrants. One family lived in a small modern house and relied on the husband's salary at a national company, which had an office in the town. After they had been settled for several years in Tosa-chō, the company subsequently insisted that the family move to a new city location, requiring them to leave the social networks they had developed and forcing the wife to find employment to afford their rent in the city. In comparison, the other migrant family had chosen to migrate to Tosa-chō to pursue a 'permaculture lifestyle' (a social movement based on sustainable and perennial agriculture); they had been motivated by a lack of trust in governmental reporting on food safety after 3.11. Despite earning little money, their education, social capital and non-reliance on corporate forms of employment made their move to Tosa-chō longer-term and more stable than that of the first family. They relied on extensive digital connections through Facebook groups and Airbnb, which brought them guests from across Japan who stayed overnight in their renovated traditional home to experience permaculture living. However, this family were also uncertain of their long-term plans, given that their parents were living far away from them and might require support further down the line.

In just one day in this small rural town, I was able to witness the wide variety of lifestyles, ambitions and different ways in which people of all ages were caring for themselves and for each other within their socio-economic confines. Many people shared a sense that they could not rely on the government, whether for adequate childcare provision, fair gendered labour practices or safe food supplies. It might be only through forms of self-sufficiency grounded in mutual support that they would be able to reduce their sense of precarity.

Aims of this book

My goal in this book is to reveal, through close ethnography, how the smartphone is implicated in the everyday lives of older people as they navigate the opportunities and challenges that come with ageing. Digital technologies, especially smartphones, are widely critiqued (both in academic literature and in popular media) for their negative effects. You only have to open a newspaper to see stories about the perils of screen addiction or the loss of 'authentic' relationships through digital communication. The mental health of adolescents worldwide has been found to be deteriorating since 2012, coinciding with the proliferation of smartphones and increased internet usage; however, causation remains unproven.[10] While such concerning observations should not be downplayed, there is room for an approach that does not only focus on the negative. This book aims to offer such an alternative perspective. It shows how the smartphone, in becoming integrated into the daily lives of people of all ages, is now entangled with practices of care. However, it is also crucial to consider why older adults are often framing their technology use against feelings of burden, and to examine how digital devices may in fact perpetuate a withdrawal of care from the state – and perhaps also from relatives.

This book considers how various dynamics, including an ageing society, internal migration, the uptake of the smartphone and the rise of visual digital communication, interact to shape experiences of ageing in Japan. The research makes several key empirical contributions. First, it focuses on the experiences of ageing of a population often neglected in studies of this kind: people in mid-to-later life, i.e. those in their mid-forties to eighties, not the young nor the 'old old'. In encompassing such a large age range, this research is positioned to question assumptions about categories of age (Chapter 2), exploring a continuum of experiences that affect individuals in different ways at different points along the later life course. Second, the book examines how the smartphone is embedded in daily life, looking at how the wider context of family dynamics, gendered labour and social norms shape technology use, along with a central question of how the smartphone is situated in relationships of care. I examine how shifting attitudes towards such things as the roles of men and women, multigenerational living, life course progression, attitudes towards health and purpose in life are impacted by and reflected in smartphone use. By including ethnographic evidence of older adults using the smartphone, we see how the device is simultaneously employed to negotiate oppressive structures within society, while both usage and non-usage replicate various societal norms. The research adopts a comparative approach across generations, between men and women and between the urban and rural. It provides a broad perspective on the parallel rise of the ageing society and digitalisation that seeks to demonstrate the variety and commonality of experiences of ageing in contemporary Japan.

My aim is to demonstrate how the smartphone is now embedded in people's everyday lives, from their engagement in paid and unpaid work in later life and the management of normative ideals of gendered labour (Chapter 3) to the ways in which people maintain social relations through acts of sharing and reciprocity on- and offline (Chapter 4). The research reveals how older adults develop visual digital repertoires as a mode of affective communication (Chapter 5), and how they deal with anxieties about health and care in the context of precarious welfare structures (Chapter 6). I also discuss how digital technologies are entangled in the movement of people to the countryside (Chapter 7) and reveal how such technologies may be part of how people respond to societal discourses that emphasise the importance of finding meaning and purpose in life (Chapter 8). The ethnographic evidence demonstrates how the later life course is imagined and lived through and with mobile communication technologies – showing how it can be both a time ripe with opportunity yet also a time of increasing responsibilities, frailties and fear of the future.

This introductory chapter presents a historical background to the lives of the people who populate this book, the majority of whom were born in the immediate post-war years. It then situates the research within the wider demographic context of Japan's ageing society, alongside the rise of smartphones and digital visual communication. I then introduce my two primary fieldsites, detailing how I gained access in both, before moving on to a brief discussion of my research methods, including a range of analogue and digital visual participatory methods. Such experiments in anthropological practice are demanded by the nature of the study, attending as it does to the rise of visual forms of daily communication now commonplace among older smartphone users in Japan.

Post-war Japan: setting the scene

In 1945, at the end of the Second World War, Japan's defeat and subsequent occupation and return to peacetime heralded an era of significant social, economic and cultural change for the country. The population spiked between 1947 and 1949 as soldiers returned to civilian life, married and had children,[11] leading to the emergence of the baby boomer generation (*dankai no sedai*), to which many of those who appear in this book belong.[12] In recent years this generation has become a focus of study for anthropologists who have documented their transition into retirement – for example revealing how caring for oneself and peers can be a marker of a 'good' retirement in a neoliberal state,[13] the dependency of retired husbands on their wives[14] and the ways in which people pursue a 'second life' (*daini no jinsei*) after retirement.[15]

The American occupation of Japan lasted for seven years, ending in 1952. Under General Douglas MacArthur, and with co-operation of the Japanese government, major political changes occurred in Japan. These included the introduction of a democratic constitution, which established many civil liberties and abolished military forces. Such changes not only had impact on a macro scale, but also attempted to affect the way in which family households were governed, albeit with limited success, as shall be seen in Chapter 2. There followed a period of rapid economic growth, the so-called 'Japanese economic miracle', founded on heavy industrialisation,[16] population mobility, productivity improvements and latent demand.[17] However, as Yoshikawa Hiroshi, one of Japan's foremost macroeconomists, argues, negative developments associated with increasing industrialisation and urbanisation included increasing pollution and the loss of rural communities and a sense of tradition.[18] Some of the implications of the latter will be discussed in this book.

Past ethnographic research provides insight into the paradoxes that characterised the post-war decades. For example, sociologist Ronald Dore's account of a Tokyo ward in 1951 documents the varied levels of deprivation of the immediate post-war years experienced by residents; his work reveals the prevalence of self- and family employment and small-scale manufacturing, while industrially produced household goods were simultaneously becoming more desirable.[19] A late 1950s study of family dynamics in Mamachi Town, east of Tokyo, by the anthropologists Suzanne and Ezra Vogel, documented the emergence of the white-collar worker – the so-called 'salary man' (*sarari man*) – who constituted a new middle class.[20] Vogel argues that the 'old middle class' (small-scale business owners) was not in a position to benefit from Japan's rapid economic growth; instead they were largely superseded by corporations with the capacity to ride out economic fluctuations. The Vogels' ethnography demonstrated how the rise of the salary man contributed to the highly gendered division of labour, with women having almost exclusive domain over the domestic sphere and men being the earners for the family, whereas previously women would have also worked alongside men in family businesses. In a similar vein, and conducted at a similar time, David Plath's (1964) ethnography on leisure time in a small city and village in Nagano Prefecture documents the experiences of the farmer, merchant and salary man and contributes to the portrayal of the attractiveness of the last as a lifestyle.

Even in rural areas, ethnographic research documented how lifestyles were changing and becoming more akin to the mainstream epitomised by the city. In sociologist Ronald Dore's portrait of the village of Shinohata, for example, he compares the lives of people in 1955, the date of his first visit, with those of his subsequent visits in the 1960s, thus documenting the post-war agricultural land reform that made farmers landowners rather than tenants of city-dwelling landowners.[21] He reveals the rapid material changes in lifestyle, as demonstrated by the proliferation of new houses, toilets, bathrooms, vehicles and appliances, while simultaneously acknowledging the fragile foundations of this prosperity, with agriculture becoming less important compared with salaried work in the cities. By the mid-1970s anthropologist Robert Smith's ethnography shows us how small-scale agriculture was rapidly becoming a pastime for older men, with the younger generations of working age having left the countryside.[22] This trend has accelerated up to the present day, with the exception of recent lifestyle migrants moving to the countryside.[23] These migrants often pursue organic agriculture and self-sustainable forms of subsistence, as outlined in Chapter 7.

The post-war decades in Japan saw a levelling of lifestyles among the population with shared 'middle-class' aspirations. However, this rising middle class did not represent homogenisation; wages remained highly varied and the standardisation of family formation (such as marriage age, family size and household nuclearisation) provoked new challenges in housing and elderly care, which further diversified ways of living.[24] Although the 'salary man' form of corporate, life-long employment grew to dominate as an ideal, other ways of earning a living remained important for community life, even in the economic boom years of the 1980s. Anthropologist Theodore Bestor demonstrates this in his ethnography of a Tokyo neighbourhood where the 'old middle class' of small business owners maintained a central position in relation to community institutions.[25]

The baby boomer generation benefited from working during the growth years of the 1960s to 1980s. However, many of those who participated in this research shared a sense that even with savings and their own home, their financial future and welfare was uncertain given the widespread economic uncertainty surrounding them. Japan's so-called 'economic miracle' peaked with the 'bubble economy' (*baburu keiki*) that expanded in the latter half of the 1980s, with staggering rises in housing and land prices. It subsequently collapsed at the beginning of the 1990s,[26] following which Japan's economy fell into severe recession and the 1990s became known as the 'lost decade' (*ushinawareta jūnen*). From the middle of the 2000s the economy began to recover, albeit very slowly. The lost decade eventually became the 'lost 20 years', with the Japanese GDP only rising by 2.6 per cent between 1997 and 2007.[27] The children of the baby boomers, who typically entered the labour force during the 1990s, became known as the 'lost generation'; they experienced insecure prospects of home ownership or stable jobs, as well as a breakdown of the male breadwinner family model.[28] Many of my older interlocutors felt sorry for their children's generation, adding to their desire not to burden them in old age.

Ethnographies of Japan over the past 20 years have revealed how such societal and economic shifts, along with the rise of neoliberalism, have created a 'gap widening society' (*kakusa shakai*) in which people live polarised lives between those who have wealth, education and good prospects and those who do not.[29] Ethnographic research has been conducted on how such societal shifts have variously affected notions of adulthood and masculinity,[30] have contributed to the precariousness of employment[31] and, of particular relevance to this book, have affected anxieties surrounding ageing,[32] challenging traditions around the performance of elderly care.[33] Contributing to a sense of precarity is the current employment situation for many Japanese people: two-fifths of the workforce are employed in 'non-regular' positions, typically earning less than full-time regular employees, with limited access to training and reduced social security benefits.[34] This is a problem that affects far more women than men, with particular implications for older women which will be discussed in Chapter 3.

As already mentioned, one of the most significant recent events in Japan was the earthquake, tsunami and subsequent meltdown at the Fukushima Daiichi Nuclear Power Plant that occurred on 11 March 2011. The latest figures reported 19,747 deaths and 6,242 injuries, mostly caused by the tsunami.[35] This 'triple disaster' has fundamentally affected many areas of life for people – including technology use, as we shall see in Chapter 5. The events of 3.11 and the aftermath have been widely studied by anthropologists. They have documented, for example, the way in which risk perception and life choices among nuclear evacuees have been impacted,[36] how the events stimulated the expansion of heritage regimes and associated culture industries in the northeast of Japan[37] and how the widespread sense of precarity that emerged after the disaster is situated within longer-term trends of economic and social decline.[38] In Chapters 7 and 8 I show how urban to rural migrants were motivated to move towards a more sustainable lifestyle because of the feelings of insecurity prompted by 3.11, and explore how they are supported in their new lifestyles by older adults in rural areas.

The intergenerational relationships that have developed serve partially to fill the generational gap that is so striking in the Japanese countryside. Migrants fulfil roles once assumed by local younger generations, who have since left to work in cities; in doing so they benefit from the expertise of older local residents while supporting them in turn. However, the lifestyles of migrant 'incomers' are often far removed from those of local people of similar age who have remained, and these two separate communities also emerge as distinct in their online interactions.

An ageing and shrinking population

Japan has a 'super-aged' population (*chō kōrei jinko*), with an estimated number of people aged 65 or older at 36.4 million, or 29 per cent of the total population – the highest among 201 countries and regions across the world.[39] This figure is expected to rise to about 40 per cent by 2060.[40] Average life expectancy reached record highs in 2020, being 87.7 years for women and 81.6 years for men.[41] An ageing population in Japan is nothing new, of course: Japanese men and women have held these records since 1985, according to the Ministry of Health, Labor and Welfare. However, what is new, and central to the motivation of this book, is the way in which the smartphone and other digital technologies have become central to how people are navigating later life.

With the combination of a persistently low birth rate and increasing longevity, the Japanese population is ageing faster than anywhere else. The situation impacts several of the more pressing issues faced by the Japanese government, from maintaining economic growth in the face of a shrinking workforce and a growing demand for pensions to providing health and social care for increasing numbers of elderly people. The rural to urban migration of baby boomers in the late twentieth century during the rapid economic growth period (discussed in more detail in Chapter 2) means that in the 2010s Japanese cities faced a rapid surge in the numbers of people reaching retirement age.[42] The ageing population ratio – the proportion of older people (typically defined as those aged 65 and above) relative to the total population – is significantly higher in rural areas compared with urban areas, precisely because of this missing generation who migrated to cities for work and had children in their new urban homes.[43] Rural areas, with their extreme proportions of the 'oldest old' residents, have become the face of super-ageing Japan, and as such represent a window to the future for the entire population. My reason for including both rural and urban sites in the design of this research was to provide a comparative perspective on the pressures facing communities and their responses to the issues mentioned above. I found that the challenges facing a depopulated and ageing remote region were precisely what galvanised it into action.

Rather than focusing solely on the problems associated with ageing, this book also considers the positive aspects of a radically increased life span. As recent bestseller lists in Japan show, there is an appetite among the Japanese population for books written by older authors about how to live well in old age,[44] demonstrating the desire for narratives that challenge discourses of decline. Among the participants in this research, expectations and experiences of the life course are changing compared with previous generations, particularly in the ambiguous time of mid-to-later-life when one is considered no longer young but not yet elderly. The wider comparative project of which this research is part has demonstrated how increased lifespans around the world have produced a period of life after the typical age of retirement; in this period continuities from adulthood are more dominant than any changes that might be expected and are often feared, associated with becoming 'elderly'.[45] This extended older adulthood presents a time when, being somewhat free of previous responsibilities such as raising a family or work, new possibilities can emerge. Yet this is not to say that responsibilities necessarily decrease in this period of life – some people experience the exact opposite, especially in terms of duties of care.

Anthropologists of ageing provide an alternative view to biomedical discourses of ageing framed mainly in terms of bodily decline, for example by focusing on 'ageing-friendly communities' responding to Covid-19[46] or by challenging the assumption that the onset of dementia and the 'fourth age' (a period of life marked by social, physical and mental decline) is synonymous with a loss of subjectivity.[47] In Chapter 2 I discuss in more depth the so-called 'fourth age imaginary', demarcated from the active 'third age' as a departure into illness and dependence.[48] Rather than focusing on people at the very end of their lives, or those in need of care to perform daily activities, this book instead documents a liminal period in which people may anticipate old age, showing how this experience blurs ageing categories.

A central question has emerged in much of the recent anthropological research on ageing: what does it mean to 'age well' and how do people navigate the challenges that may come with advancing years while balancing autonomy and dependency?[49] As I discuss in this book, for many of my research participants 'ageing well' was framed in terms of maintaining a sense that they were staving off any burden that they might cause other people – that they were capable of taking care of themselves. Digital technologies are emerging as a key way in which people are taking care of themselves and others. However, this framing of ageing individuals as burdensome, and the focus on individual responsibility, requires unpacking and contextualising, particularly given the precarious conditions of neoliberal capitalism in which many Japanese find themselves today. Older people often endeavoured to remain socially embedded through the act of caring for others and contributing to their communities, especially when they might need increased care themselves. Many of my participants felt they could contribute by helping to care for still older people, filling gaps left by weakening state welfare and a diminishing birth rate.

Recent research has shown the wide diversity of culturally inflected paradigms of how people 'age well', critiquing assumptions that 'successful ageing' requires independence, activity and agelessness, as is normative in many industrialised nations.[50] Anthropologist Iza Kavedžija, for instance, has demonstrated how in Japan maintaining independence in later life requires cultivating dependences on others.[51] Similarly, the work of anthropologist Sarah Lamb demonstrates the value of interdependence and transience in later life in India.[52] Lamb challenges cultural assumptions underpinning dominant discourses around ageing that 'overemphasise independence, prolonging life and declining to decline'.[53] In Kavedžija's work among older adults in Osaka, Japan, it is precisely later life that grants her research participants liberty after fulfilling social roles such as being 'a good wife and mother', yet these new possibilities are experienced simultaneously with various forms of decline. This book is inspired by Kavedžija's approach towards ageing that acknowledges the entanglement of decline with new possibilities in later life.

The rise of smartphones

The rise of the adoption of smartphones among older people in many countries around the world means that they may now be regarded as devices for everyone, not only the young.[54] In Japan the adoption of smartphones has been slower than in other parts of the world, in large part due to the pre-existence of internet-enabled feature phones (*garakei*) that previously dominated the market, the impact of which will be outlined in Chapter 5.

While smartphone usage is one focus of this book, the ethnography highlights how people situate their smartphone use in already existing digital and non-digital practices, for example of care, in daily life. The intention is to appraise the impact of the uptake of smartphones by older adults, while acknowledging that a vast amount of life for this population takes place outside of the smartphone. Indeed, several research participants, both with and without smartphones, rejected the device on the grounds that it was not necessary, that it was too expensive or that it was a device devoid of 'heart' (*kokoro*). However, a close ethnography pays attention to what people *do* as well as what they say, and in many cases people's ambivalent appraisal of the smartphone, if they did indeed own one, was not supported by their actual usage of the device to connect with others, sometimes forming the centre (*chūshin*) of their social lives. It is not the intention of this book to overstate the significance of smartphones in the lives of Japanese older adults, but rather to provide a snapshot of a moment in which they are being taken up by this population, to document their reasons for doing so and to analyse the centrality of care via the smartphone in a wider context of precarity and uncertainty surrounding ageing.

While young people may have originally shaped mobile culture,[55] older people now make up a large proportion of mobile phone users and are shaping smartphone culture in their own ways. Mobile phones themselves are also adapting to use by older people. For example, popular with several of my older research participants are the so-called 'easy (*raku raku*) smartphones' developed by a number of Japanese technology companies, including SoftBank,[56] Fujitsu and NTT Docomo.[57] These phones are equipped with useful features such as magnifying software and an easily navigable interface with big icons and text. Sato-san, a 90-year-old teacher of flower arranging (*ikebana*) in Kyoto, uses an easy smartphone. While it had not been her decision to purchase the device, she soon adopted it with enthusiasm:

> I always had a *garakei*. And then, my niece, she's 21 years younger than me, she was changing hers to a smartphone and was like, 'right, we're changing yours too!' [laughs]. So I was more forced to than a choice! But it's really convenient! I can look up the weather, and see if it's going to rain in my area.... Just type in Kitano. And this!! This is so convenient... LINE [shows the application LINE]. I chat with my students and they're all working, right? So I never know what time it suits them best. So I use this to talk to them.

From my observations of older adults, and discussions with them about their use of *garakei* and smartphones, it was clear that while *garakei* are internet-enabled, the adoption of the smartphone afforded many more capabilities than were possible on their older devices. These included more opportunities for personalisation, evident in the exterior and interior of the device. Many of the affordances were visual in nature. For example, while *garakei* typically had integral cameras, the resolution was low and difficulties could arise in sending and receiving photos between smartphones and *garakei*. People who made the switch to smartphones were thus able to participate in more visual forms of communication through the sharing of photographs. They were also able to use the messaging application LINE, which represented a radical change in their mobile phone usage and possibilities of communication.

Visual digital communication

One of the aims of this book is to document fundamental shifts in human communication brought about by the rise of visual digital media sharing via the smartphone. Social media platforms such as LINE, WhatsApp, Facebook and Instagram have afforded the sharing of visual elements, for example photographs, memes, emojis and gifs, which can now be considered as important for communication as speech and text.[58] This is a global phenomenon of considerable significance,[59] but it also has local and national trajectories. With its history of rich visual culture, Japan offers an ideal opportunity for exploring the implications of this fundamental shift towards the visual with respect to a particular population. This book demonstrates how digital images are now so integral to human sociality that they have also become integral to anthropology.[60]

Visual anthropologists have long been concerned with the study of visual material and practices, comprising forms of representation of the people among whom we research[61] and constituting a way of seeing[62] and sensing[63] within ethnographic practice. Much of this early research concerned photography, followed by video. In the years since the proliferation of camera phones, research has turned to the sharing of digital visual media, assessing how it impacts on conceptions of public and private spaces[64] and exploring its role in the maintenance of social relations.[65] Ethnographic research conducted in Japan reveals how the rise of camera phones enables photo-sharing as a form of 'intimate virtual co-presence' among young people.[66] More recent research has shown how visual messaging via the smartphone is deployed intergenerationally to maintain family ties across geographical distances.[67]

This book builds on these discussions towards an appreciation of how the sharing of visual digital media has become commonplace among older adults. The visual media discussed comprises photos, videos and, most distinctively in the Japanese context, an abundance of illustrated messages. In Chapters 2, 3 and 5, I argue that illustrated messages in the form of 'stickers' (*sutanpu*) (large-format emojis that are sent as individual messages) perform several key functions which have resulted in their rapid uptake by older adults. First, they can be considered as 'affective objects' that simultaneously replicate and challenge gendered labour norms (Chapter 5). I employ the concept of 'emotion work', which refers to the effort individuals intentionally make to induce or inhibit feelings so that they are appropriate to a given situation.[68] As anthropologist Iza Kavedžija has argued, communication in Japan, with its attention to politeness and concern for the correct way of doing things (*chanto suru*), demonstrates the care that someone is showing to another person, rather than indicating a lack of intimacy.[69] In this way, stickers enable people to convey the 'correct' feeling of a message, managing their public facade (*tatemae*) through the careful selection of digital visual media. For older adults who may not be confident with using a small smartphone keyboard, and for whom crafting textual messages could take a significant amount of time, stickers and emojis are particularly convenient and efficient markers of affectivity; they can also be highly personalised.

I draw on the notion of phatic communication, coined by the anthropologist Bronislaw Malinowski in the 1920s,[70] to show how in the Japanese digital context stickers are used in place of expressions valued for their sociality rather than their inherent meaning. The choice of a particular sticker corresponds to a careful alignment of personal representation and the performance of self,[71] embodied communication and communicative fluidity[72] – all significant for a society in which the form that communication takes is just as important as the message itself.[73] Moreover, visual media, such as stickers in the Japanese context, are used to create an amiable 'atmosphere' to maintain harmonious relationships; they draw upon the Japanese communicative custom of 'reading the air' (*kuuki wo yomu*) and build on recent work that shows how ambient media consisting of video, music and text creates 'atmospheres of self' in Japan.[74] I propose graphic methods as a useful way for anthropology to approach the study of how affective experiences and digital practices intersect.

The possibilities that arise from this fundamental shift towards the visual, both in human communication and in anthropology, are further explored in this book by looking at how the smartphone has become a tool through which people manage their own visibility.[75] I show how older adults craft visual digital communication that aligns with their interests and personalities, affording them digital visibility that might not be possible in their offline lives (Chapter 4). Visibility and invisibility are discussed in Chapter 2 in relation to shifting domestic arrangements and the desire for privacy, both of which are mediated by technology use in physical and digital domestic spaces. On the one hand, the smartphone gives people increased control over their visibility through what I call 'tactical invisibility', reducing the oppressive scrutiny arising from multigenerational living that burdened previous generations. On the other hand, the very same device is also a tool for surveillance; people are equally aware of the 'pressure' that comes from visibility through the smartphone (Chapter 4). Foucault famously wrote that 'visibility is a trap',[76] yet I argue that this visibility enables care at a distance and facilitates a lowering of the burden of sociality (Chapter 5). In Chapter 6 I show how the social gaze can be a key motivator for behaviour change in a digital health intervention that involves visual digital media. The significance of the smartphone in the fine line between care and surveillance was observed during my in-situ fieldwork. It intensified further during the Covid-19 pandemic, when private online spaces were co-opted by the Japanese government in their pandemic response – a development that I argue was particularly challenging in a context where people are already highly sensitive to public scrutiny (Chapter 6).

I analyse digital visual material to show how the sharing of digital photos relates to the traditional forms of sharing (*osusowake*) and ideas of reciprocity (*yui*) that exist, especially in rural Japan (Chapter 4), broadening anthropological discussions of digital sharing centred on self-representation.[77] In Chapter 6, focusing on health and care, I situate current digital health tracking practices, which rely on visual markers and rewards, within historical tendencies towards visual self-tracking associated with Shinto pilgrimage. Here I ask how digital health interventions can be more usefully aligned with the everyday digital practices of older adults that involve visual digital media, rather than relying on bespoke health applications. In the penultimate chapter on life purpose, I return to earlier discussions of authenticity and visibility to analyse how people experience purpose in life (*ikigai*) not as an externally mandated category, but rather as something personal and meaningful that emerges particularly in challenging life circumstances and is mediated through visual digital modes of self-reflection and sociality. Finally, through studying these trajectories of visual culture and their significance to older populations, this book aims to contribute to an anthropological understanding of the good life and happiness.[78] In doing so it explores the complex relationship between 'living well' and 'life purpose'[79] and, more broadly, situates digital practices within the emerging area of existential anthropology.[80]

Fieldsites

Informing this book is 16 months of ethnographic research, living with and alongside my research participants, conducted primarily in two fieldsites between February 2018 and May 2019 (Fig. 1.1). For the duration of the fieldwork I was based in the Kansai region of Western Japan, beginning with two months staying in a family home in Osaka before moving to neighbouring Kyoto. Six months into fieldwork I began research in a third site – a rural town called Tosa-chō in Kōchi Prefecture in Shikoku, southwestern Japan – which I would visit every month, typically for up to ten days at a time. I also followed several of my participants as they relocated to other parts of Japan, including to Tokyo.

During my stay in Northern Osaka I was immersed in the Katsura family household, staying in a bedroom that had previously been occupied by one of their sons. This multi-generation home housed a couple in their mid-sixties and the husband's parents, who lived in the apartment on the ground floor. Chapter 2 will detail the layout of the Katsuras'

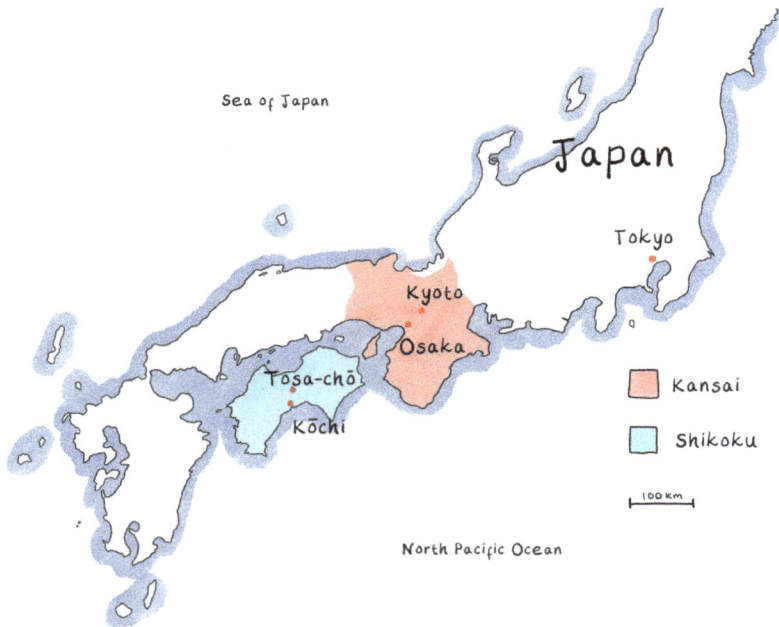

Figure 1.1 Map of central and southwestern Japan highlighting the regions and locations in which the research was conducted. Image by Laura Haapio-Kirk.

'two generational' home, and the background to this kind of dwelling. The two households were quite separate and I would eat meals only with the middle-aged couple. Staying here offered an opportunity to conduct deep ethnography in a Japanese home – a private space into which visitors are rarely invited (Daniels 2010). Chapter 2 details the kinds of care I witnessed within and between the two households, showing how the kinds of intergenerational care that are embedded in material practices in the home are shifting to digital practices that emerge in the study of digital communication with their children.

The area immediately around where I lived and conducted research in central Kyoto, which I shall call Kitano, is a mix of residential buildings, schools, shops, restaurants and a few remaining craft workshops (Fig. 1.2). These included a tea ceremony water pot maker, a knife maker and a lantern maker who had been in operation for more than 300 years – which for Kyoto, where families take pride in tracing their rooted ancestry back generations, is just about enough to be able to call yourself 'local'. Kyoto is known to be the centre of many Japanese cultural traditions. People

Figure 1.2 A short video introducing the Kyoto fieldsite. https://youtu.be/ONlL4EuBSTc.

come to visit from all over Japan, and the world, to learn about the tea ceremony, traditional dance and music, the world of geishas (*geiko*), flower arranging (*ikebana*), silk weaving, paper crafts, etc. Precisely because of this intense concentration of cultural and artistic activity, many ethnographic studies of Kyoto have focused on such practices: those of silk weavers,[81] the traditions of geisha,[82] and the ways in which city planning and tangible cultural heritage shapes daily life for Kyoto residents.[83] Anthropologist Jason Danely has conducted research on memorialisation and loss among older adults in the city.[84] Anthropologist Eyal Ben-Ari has conducted research among mainly older adults participating in neighbourhood associations in two suburban areas close to Kyoto.[85]

In Kyoto city centre it is typical to see people of all ages: older women pulling shopping trolleys, young women on 'mama bikes' (*mama chari*) with their children on the back seat, office workers in suits grabbing a quick lunch from convenience stores, older men reading newspapers at cafes and, of course, plenty of tourists – many dressed in rented kimonos – on the main shopping streets, at temples and in the more historic streets lined with restaurants and tea ceremony houses dedicated to serving them. Walking through the streets of central Kyoto, the population that you encounter does appear very mixed, masking the ageing population that defines twenty-first-century Japan. The more time one spends here, however, the more visible that ageing population becomes. A magazine rack at the entrance of a cafe holds not only newspapers, magazines and manga; it also features a collection of reading glasses of different strengths for customers to borrow. The ATM machine at the convenience store has a holder in which a customer can place their walking stick while they withdraw cash. The uniformed city workers, employed to direct pedestrians around building works by waving brightly coloured pointers, for example, are typically all men in their sixties and seventies.

To protect the anonymity of the people in this research, I do not refer to the names of local temples or other identifying landmarks as these would allow a reader to know the exact location of my research. However it is important to state the general location of this city centre fieldsite as it was undergoing such rapid change, compared to what I could see in other parts of the city. The ward in which I lived is called Shimogyō-ku. It covers the urban core around Kyoto Station, stretching up to Shijo dori, the busiest shopping street in the city – especially during the spring cherry blossom season (Fig. 1.3). The surrounding streets retain a surprising sense of calm and quiet. The narrow streets typically only have space for single lane traffic; they are lined with a mix of modern structures and two-storey old wooden houses (*machiya*), now often converted into hotels and restaurants.

Over the course of my fieldwork I witnessed the rapid change taking place in this area. New hotels were springing up in a matter of months, while old businesses, often owned by people in their seventies, were being replaced by trendy boutique coffee shops. Residents told me how the area had dramatically changed over the past 60 years, noting that many family-run businesses had vanished – driven away by rising land tax prices. The rapid changes I witnessed will be detailed in Chapter 4 through the accounts of several long-term residents, focusing on the reduction in public spaces in which older people can comfortably spend time and socialise. Such a reduction, I will argue, can negatively affect older men in particular.

Although I was based in one neighbourhood, I inevitably made connections with people across the city as my network expanded. However, the area in which I lived formed the basis of community interaction and demonstrated the importance of institutions such as temples in community life. A typical day in Kitano begins at 6 a.m. with the ringing of Buddhist temple bells, which can be heard from within most homes given the permeability of Japanese buildings. The local temple near to my apartment emerged as a central site for the community, hosting various weekly activities and events throughout the year. Every street that runs east to west in that area of Kyoto is named after a temple on that street. Many small Shinto and Buddhist shrines are dotted around, typically maintained by older local residents and used during community

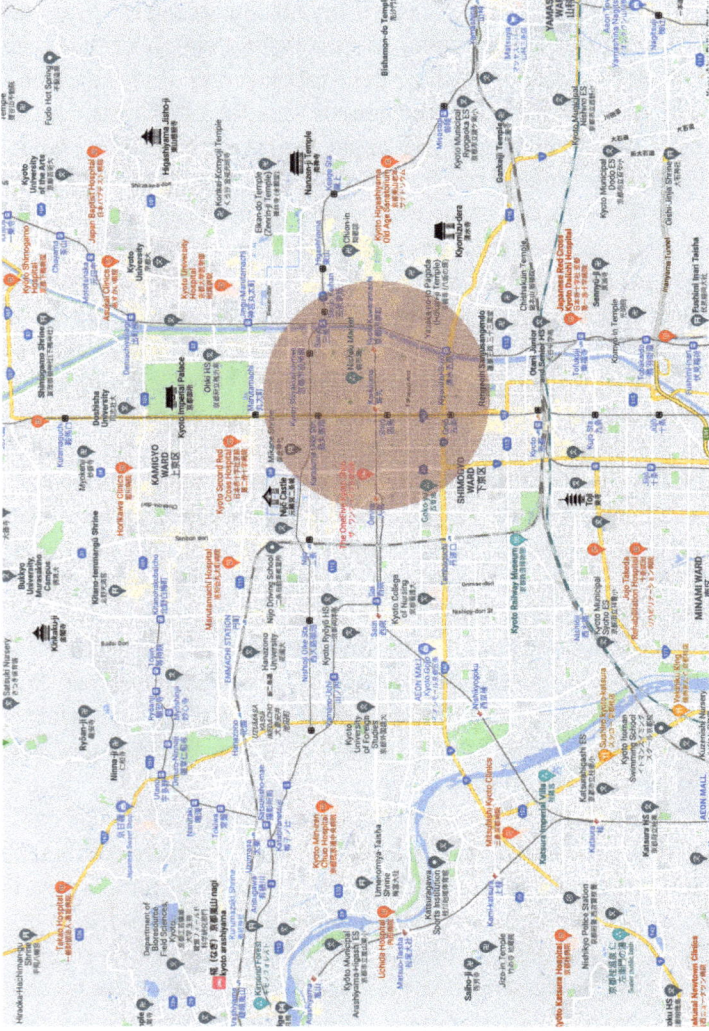

Figure 1.3 The Kyoto fieldsite in the centre of the city near the intersection of Kawaramachi-dori and Shijo-dori – two of the main roads that run the length and breadth of Kyoto. Map data ©2021 Google.

festivals such as *Jizobon* in August. Even though most people I spoke with proclaimed themselves not to be religious, it appeared that everyone participated in some way in traditions associated with religion and Kyoto-specific festivals such as the *Gion Matsuri*. Such festivals throughout the year emerged as highly visible moments of neighbourhood sociality, conspicuous in a community where during daily life people tended to keep to themselves. The visible and more hidden forms of community that I became aware of during my fieldwork will be discussed in Chapter 4 when I compare physical and online sites of sociality.

My rural research was conducted in Shikoku, one of the larger islands that constitute Japan, located in the southwest of the country. Shikoku is remote from the urban centres of power situated in the Kantō and Kansai regions, and people in Kyoto were surprised at how often I was going there, an area that many people had never visited themselves. The Reihoku region in northern Kōchi Prefecture, where I was based, is especially remote, dominated by forested mountain ranges within which are nestled small towns, villages and hamlets. The Reihoku area is located about an hour's drive north of Kōchi city in the centre of Shikoku (Fig. 1.4). Residents would either drive via a new toll road expressway, which goes directly through tunnels carved out of the mountains or (if they did not want to pay the toll) take the longer route via a road that winds its way over the steep mountains separating the region from Kōchi city. This drive through sparse settlements and abandoned mountain villages emphasises the remoteness of the location.

Figure 1.4 A short video introducing the rural fieldsite in Kōchi Prefecture. https://youtu.be/HC_XBr0-D8g.

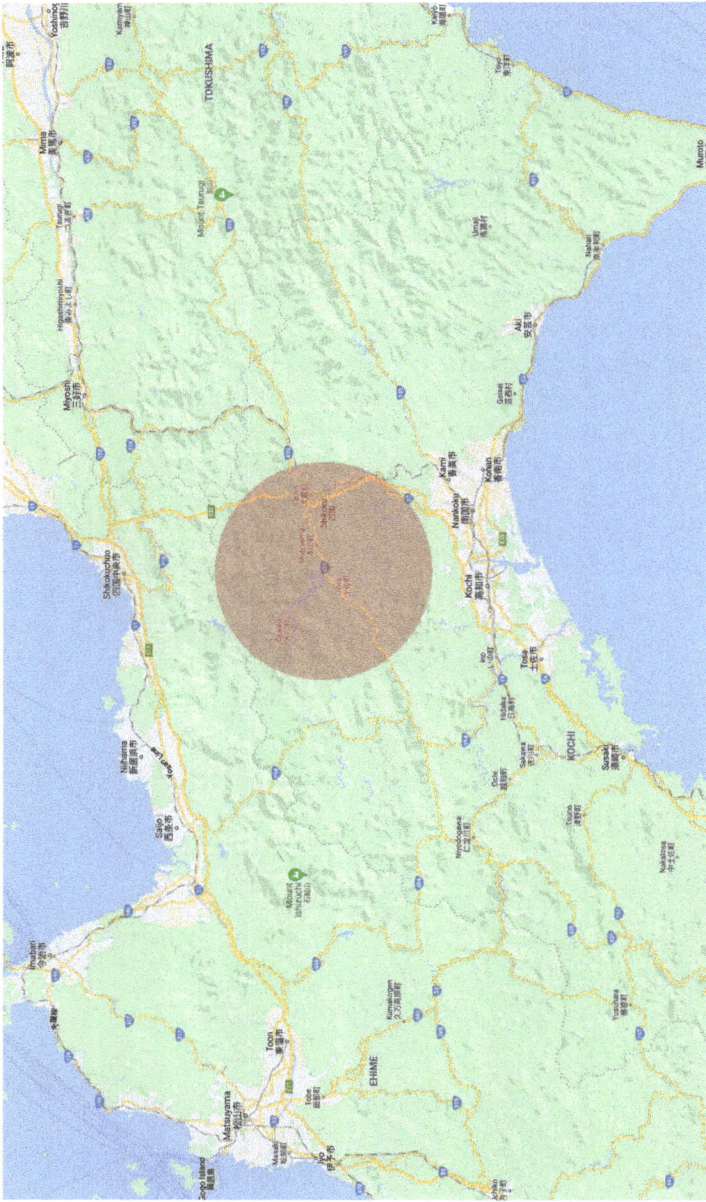

Figure 1.5 A map of central Shikoku, highlighting the rural fieldsite Tosa-chō in the Reihoku region of Northern Kōchi Prefecture. Map data ©2021 Google.

The Reihoku area comprises about 757 km² and has three towns (Otoyo-chō, Motoyama-chō, Tosa-chō) and one village (Okawa-mura) (Fig. 1.5). During my research visits I predominantly stayed in the town of Tosa-chō. However the distinction between Tosa-chō and neighbouring Motoyama-chō is not clear, with many people living and working between the two. My research participants included people from all three towns and smaller villages. In order to protect anonymity in these small communities I simply refer to participants as being 'from Tosa-chō' throughout this book, rather than specifying their location. I also choose to focus on Tosa-chō because this is the town in which I resided during my fieldwork and developed stronger connections with local government, health and social workers.

While the populations of both Tosa-chō and Motoyama-chō predominantly reside along the central Yoshino valley, where the hospitals, schools and supermarkets are located, they are not really 'towns' in the sense of being a clearly defined area of habitation with one central point. Rather, they are better understood as wide-ranging municipalities. For example, Tosa-chō is organised into 15 districts spread over a large area of 212 km², of which 87 per cent is forest[86] (compared to the national average of 67 per cent).[87] Only 1.7 per cent of Tosa-chō is agricultural land and 0.6 per cent is residential, thus representing a typical Japanese mountain village landscape.[88] This geographical spread poses problems of potential isolation for the ageing population as well as prompting innovative solutions by the local government, which will be discussed in Chapter 7. Frequent typhoons occur in the region, accelerated by climate change, and the mountainous topography makes the area vulnerable to landslides which have previously cut off access to villages.[89]

Japan is one of the most urbanised countries in the world, with 79 per cent of the population living in urban areas.[90] One may therefore question my decision to include a comparative rural fieldsite in this research. One of my reasons was that rural Japan, with its extreme ageing population, represents the 'face' of super-ageing Japan. In Tosa-chō, 45 per cent of the population are aged over 65,[91] compared with the national average of 28 per cent.[92] The national population of Japan is predicted to reach similar proportions by 2065.[93] The OECD defines a 'super-aged society' as one in which more than 21 per cent of the total population is aged 65 years and older.[94] In the next 30 years the speed of ageing in Asia-Pacific countries will be unprecedented, with 11 countries being predicted to reach 'super-ageing' status.[95] Rural regions such as Tosa-chō therefore offer glimpses into the demographic future not only of Japan, but of much of the world.

While statistics such as these offer some indication of what the future holds at a population and societal level, only by spending time in rural areas can one appreciate how these demographic shifts affect everyday life for individuals across generations. I found that home-based health and care services for the elderly, such as home visits from doctors and social workers, are of particular importance in this rural area, especially when elderly people were living alone, for example after the death of a spouse. The children and grandchildren of the older adults I came to know typically lived in cities. Push factors for their migration included scarcity of rural employment, while pull factors included better paying work and education opportunities. The promise of higher status, convenience, modernity and 'coolness' have also been found to draw rural migrants to Japanese cities.[96]

Equally interesting are the myriad reasons why people return to their rural roots, however, and how technologies such as the smartphone enable them to make the transition back into rural life more easily. I found that the smartphone is particularly important for so-called 'U-turners' – older adults who have retired back to their hometown. Smartphones help to make rural life more convenient and enable relationships of care to be maintained at a distance, as discussed in Chapter 7. My rural-urban comparison facilitated an understanding of wellbeing as construed by social relationships and mutual support. Building on anthropological work by Traphagan[97] in rural Japan on the link between the wellbeing of the group and individual, my rural research focuses on individual health as a kind of social capital which binds the community.

Research participants

> Yeah, well actually, 80 is still so young! There are so many people who are youthful at the age of 90. When 99 hits, it's like 'Oh! Are they about to age?!'
>
> Care manager at a residential home for elderly
> people in Kyoto

To understand the use of smartphones by older adults it was important to get a sense of how people from a wide range of ages use smartphones and think about ageing. This research focuses on people typically between the ages of 45 and 80, but I also include people outside of this wide range. Given the length of life expectancy in Japan, there is considerable variation in how people experience ageing; the frailties of old age can come on much later in life than expected, prolonging middle age or even, in the words of the care manager above, youth.

Anxiety about ageing has been found to peak for Japanese people in their forties,[98] a time when they are most likely to be in full-time employment and therefore shouldering the burden of an ageing society, hence my decision to conduct participant observation with people at, and above, this age. Those aged between 45 and 49 report worries about financial dependency in later life and exhibit the lowest levels of life satisfaction and happiness of all age groups.[99] Yet much of the literature on ageing in Japan tends to focus on only the oldest section of the population. This midlife group, by contrast, remain under-studied from the perspective of ageing and from the literature on mobile phones, which tends towards studying appropriation by young people. This research documents how mid-to-later life is changing, as increasing longevity combines with new technological capabilities that can facilitate experiences more typically associated with youth.

Because of the highly gendered nature of socialising in Japan, most of my closest research participants were female. I did develop friendly and supportive relationships with several male participants, including those who I would regularly meet individually and as part of a couple, but I did not feel they were as open with me as my female research participants were. I sought to balance this gender bias by conducting semi-structured interviews with more men than women. In my rural fieldsite many of the local participants were in their sixties, seventies and eighties. It was more difficult to spend time with younger local people, who were often working several jobs, for example in restaurants and stores; there were also fewer younger residents compared with older generations.

I also developed deeper friendships with urban to rural migrants because of the kinds of self-employed jobs they had, giving them more time to socialise compared to the local people. Many of this group had also travelled internationally and were keen to develop relationships with me. I sought to remedy this imbalance through repeat visits to the homes of local older participants, who were always extremely welcoming and had more time to engage with me, and by visiting local people of working age in their workplaces, such as two local childcare facilities, restaurants and cafes. Numerous relationships were built through these strategies, leading to further contact with younger and middle-aged locals.

All the names of my participants[100] have been anonymised through the use of pseudonyms, except in the case of participants who wished to be named – for example, Ito Megumi whose painting appears in Chapter 8. Megumi-san is a practising artist; it was thus important that her contribution to this book and subsequent publications be recorded in association with her name. I also pixelated the faces of participants who wished to remain anonymous in images or videos.

Participant observation and digital ethnography

My primary research method was participant observation, which allowed me to examine the interplay of offline and online practices in relation to experiences of ageing.[101] Participant observation is the typical fieldwork method for anthropologists, usually involving simply 'hanging out' and engaging in daily activities with people (Fig. 1.6). The ubiquity of the smartphone in popular discourse – and in daily life for many of the people I worked with – presents both opportunities and challenges for anthropological research. Many people had an ambiguous relationship with the device; on discovering what my research was about, they would often assume that I was researching negative effects of smartphone use, from radiation concerns to online bullying, both topics of media attention at the time. When asking them directly about the smartphone, they would therefore admit that it was 'convenient' but that they might be spending too much time on it; they tended to be dismissive and brief in discussing it. To approach the smartphone directly as a topic of investigation

Figure 1.6   A short video describing my fieldwork methods. https://youtu.be/i9IftSx73I4.

I conducted 25 smartphone-based elicitation interviews, during which I asked participants to show me all of the apps on their phone.

While interviews were helpful to see the range of apps people had on their devices, there turned out to be only a limited number of apps that people actually used. Therefore in several instances I prioritised allowing people to spend more time discussing the few apps they did use in depth over going through the entire phone, app by app. This kind of object-elicitation interview, based on the technique of photo elicitation,[102] presented moments to discuss a range of topics tangentially related to the smartphone. I encouraged participants to expand on areas that they seemed eager to approach, believing this to be more valuable than focusing narrowly on the device in their hands. Such a conviction reflects my wider approach to this ethnography.

The smartphone emerged as a key tool during fieldwork, both for facilitating social connections with people and for conducting ethnography. I found that it was in engaging with people via the smartphone, rather than the smartphone interviews themselves, that provided most insight into how the device figured into their lives. Often people were eager to swap LINE contact details upon meeting; over the course of fieldwork I acquired 111 LINE contacts and became a member of 21 group chats. This connectivity also applied to other social media. I am connected with similar numbers of people across Facebook and Instagram, where I also follow local community accounts such as that of the town hall in Tosa-chō and local businesses.

Interacting with people on these platforms was integral for embedding myself in relationships with individuals and within communities, and such online interactions continued once I left Japan. In a digital era, the notion of the fieldsite is challenged.[103] One never really leaves the field, therefore these platforms have enabled valuable opportunities for continued connection, friendship and collaborative multimodal research once I returned.

My iPhone was my primary camera and voice recorder in the field. There was seldom a day on which I did not photograph or video something, and it became a visual fieldwork diary to which I still often turn, in addition to my written fieldnotes. The metadata such as location, date and time, embedded in digital images, makes for highly searchable and useful records of the entirety of my fieldwork experience. I often shared photos with my research participants of our outings together and they would do the same, turning the digital media we produced into valuable materials of social exchange. The smartphone was equally, if not more, integral to my remote fieldwork, which continued after I had returned to the UK.

My research significantly benefited from assistance by Sasaki Lise, a Japanese medical anthropologist, who accompanied me on all visits to my rural site, translating and transcribing the semi-structured interviews recorded during the research. Compared with my research conducted in Osaka and Kyoto, based primarily on one-to-one encounters with individuals, the research in rural Kōchi was more interview-based. One of the greatest advantages of having Sasaki accompany me on these trips was the opportunity to discuss the research together with someone who knew these people as well as I did and to spot emerging themes together – in addition to sharing the challenges and joys of fieldwork with someone who became a good friend. We have produced several co-authored articles based on this research, published elsewhere.[104]

Multimodal ethnography

This book delves ethnographically into visual modes of digital communication. It also explores how anthropologists might appropriate graphic forms to make the discipline more explicitly collaborative and accessible.[105] Through a range of visual methods, including photography, film, drawing, painting and collaboratively produced graphic narratives, this book uses visual methods as a mode of enquiry into increasingly visual modes of relating. The drawings at the start of each chapter offer a visual meditation on the stories to follow, combining fieldwork sketches with illustration as a tool for thinking. I discuss the analytical merits of drawing in Chapter 9. In my research, the smartphone itself emerged as a central tool for digital visual ethnography, including for remote collaborative methods. Ethnographic fieldwork is now often conducted via and even inside digital technologies, many of which are highly visual in nature.[106] From Google Maps to social media to the camera app, my smartphone became an essential tool for exploring and participating in my fieldsites, as is the case with much recent anthropological work.[107]

Upon discovering the importance of digital visual media through observing the way in which my research participants were using their smartphones, I foregrounded experimentation with digital visual collaborative methods.[108] As anthropologists Favero and Theunissen note, the prevalence of smartphones and other digital devices means that researchers, their audience and their participants increasingly share access to the same means of representation and technological skills.[109] This digital-visual 'moment'[110] presents anthropologists with the opportunity to commit more fully to collaborative research practices that can endeavour to balance dynamics between researcher and research participants, and so contribute to a more egalitarian discipline – for example, in the case of collaboratively produced digital exhibitions.[111]

Because of my findings on the importance of visual modes of relating to the world and the prominence of manga as a cultural medium in Japan, in addition to my personal interest in drawing, I employed graphic research methods by engaging participants in drawing exercises (Chapters 4 and 5) and in co-scripting comics (Chapter 9) that narrate their experiences of the smartphone.[112] The use of drawing in anthropology declined with the separation of the physical and social sub-fields of the discipline, and is only now undergoing a resurgence.[113] Through experimentation with digital multimodal approaches, this book contributes to visual anthropological arguments that challenge the centrality of text in ethnography,[114] and to a growing interest in developing drawing as a research practice.[115] In the co-production of multimodal material with my participants, my hope is to produce research that is also meaningful and accessible to the people with whom I have worked.[116]

Participatory visual methods put the research participant's experiences and ideas at the centre of meaning making.[117] They open up possibilities for a more egalitarian relationship between researcher and participant, with the resulting visual narrative a direct result of their collaboration. However, multimodal methodologies are not removed from being complicit in the power structures and hierarchies of knowledge production that have characterised the discipline of anthropology.[118] In the co-creation of visual research and dissemination materials, my intention is to enable participants to determine the framing of their experiences in personalised ways.[119] However, I am also aware of the inherent biases around what 'counts' as scholarship that influence how this material will be received. I also include a selection of my own field sketches and graphic narratives produced throughout fieldwork and afterwards, which served as valuable material to develop my analysis alongside written fieldnotes. As the anthropologist Elizabeth Hallam has suggested, 'rather than forming separate processes, writing and imaging are implicated and interdependent in the making and use of the field-sketch'.[120] I would argue that these are also interdependent in their analytical offerings.

I suggest that by combining digital ethnography with graphic methods we might better understand how digital repertoires[121] are enmeshed in everyday sensory, spatial and temporal practices. As anthropologists Ardèvol and Gómez-Cruz assert, digital ethnographic methodologies may be based on long-standing forms of ethnographic practice, but they represent new ways of knowledge making and thus may lead to new kinds of knowledge.[122] I would add that graphic methods, such as the drawing of maps and diagrams, have a long-standing history in anthropology despite their subordination to text.[123] Now, however, with the proliferation of visual digital media in everyday life, they open new ways of knowing about how such media practices influence how we experience the world. Furthermore, in the production of graphic forms of dissemination such as comics and illustration, we might further the sharing of such knowledge so that the people who contribute to our research may also gain access to it.

Drawing, which was part of the early ethnographer's toolkit, from sketches of material culture to kinship diagrams, is now making a resurgence in the guise of a recent 'graphic narrative turn' in anthropology.[124] Participatory drawing as a research method has been mostly employed in the study of children.[125] However, digital visual methods, such as photography and video, have been far more widely appropriated.[126] In Chapter 5 I discuss drawings made by research participants depicting their relationship to their smartphone, Chapter 4 presents participants' drawings illustrating the concepts of private voice and public facade (*honne and tatemae*), while Chapter 8 presents their drawings and a painting reflecting ideas of purpose in life (*ikigai*). In foregrounding visual participatory methods in the study of digital practices, my research seeks to highlight subjectivities and emotions, and to facilitate the co-construction of knowledge.[127]

Ethnography has traditionally prioritised the ethnographers' vision and hearing (Hockey and Forsey 2012), while relegating the other senses and neglecting an integrated corporeal sense of being in the world. The work of sensory anthropologists, such as Sarah Pink[128] and Paul Stoller,[129] challenges ethnographers to go beyond the remit of sight and sound. I suggest that co-created comics are one way of conducting multimodal research that enables research into – and the dissemination of – people's embodied, multisensory digital realities. The comic scripting process, as discussed in Chapter 9, situates the smartphone user within a participant's physical environment, while simultaneously allowing access to interior monologues and to the private spaces of their smartphone. In doing so, we gain a sense of a person as enmeshed simultaneously in multiple online and offline relationships and realities.

Conclusion

In studies of the internet and new media, there is a persistent concern about the issue of technology-mediated humanity. The underlying fear has been that the ubiquitous appropriation of digital communication technologies would make human beings lose some of their 'real' selves and lead to a decline in 'genuine' relationships.[130] Yet this book accords with the assertion of Dorinne Kondo, an anthropologist of Japan, that people have multiple selves[131] and aims to contribute to a growing body of research arguing that offline relationships are no more authentic than online ones[132] – indeed that online lives can sometimes be more 'real' in terms of personal aspirations and the feeling of 'being-at-home' than those offline.[133]

This book reveals how the dramatic lengthening of the human lifespan, along with the adoption of smartphones among older people, has for many people opened up new possibilities of what later life can be.[134] However, ageing populations also represent one of the greatest challenges of this century,[135] prompting urgent research in a wide range of academic disciplines, sectors and industries to understand and mitigate some of the societal and economic burdens of ageing. A move towards the digitalisation of public services such as health and welfare is one way in which governments are seeking to reduce their spending as people rely more heavily on such services for longer proportions of their lifespans.[136] It is therefore essential to understand the culturally and socially inflected ways in which older people use digital technologies in their daily lives, or do not, in order to ensure that digitalisation is inclusive and effective for all segments of the population.

In the following chapters, the daily realities of peoples' lives are revealed through the stories of older people as they adopt digital technologies. Participant observation allows an anthropologist to closely examine experiences that may otherwise be invisible to other disciplinary approaches. In witnessing the ways in which the daily practices of older adults respond to the wider context of societal ageing, and the various forms of precarity that it can engender, anthropologists can simultaneously observe how people creatively navigate the challenges and opportunities of later life to define their own experience of ageing.

Notes

1. Moore and Atherton 2020.
2. Throughout this book the English plural form 'emojis' will be used instead of 'emoji', which in Japanese refers to both the singular and plural.
3. Vogel 2020 [1963].
4. Emperor Akihito abdicated, marking the end of three decades of the *Heisei* era under his reign and the start of a new era called *Reiwa*, under his son Naruhito.
5. Baseel 2019.

6. Chieno-san wished to be named in this research.
7. Anthropologist Iza Kavedžija (2019) notes how repeated stories indicate their importance.
8. The term 'precarity' will appear throughout this book, referring to a condition of precariousness or instability, often related to employment, housing, healthcare or other state-based factors. The concept of precarity has gained prominence in the social sciences (Millar 2017), including in anthropology (Han 2018), to describe how people navigate contemporary economic and social challenges around the world (for example in Japan see Allison 2013).
9. Rural municipalities have both 'U-turn' and 'I-turn' policies to attract two kinds of domestic migrants (*ijū-sha*) to shrinking towns and villages: those who were born in rural areas, have worked in cities and are now returning to their rural hometowns (so-called 'U-turners') and those who were born in cities and decide to relocate in a one-way movement to the countryside (so-called 'I-turners').
10. Twenge et al. 2021.
11. Kazuo 2015.
12. This book considers the baby boomer generation to include those born between 1947 and 1951, which is not uncommon in the literature (Asao 2007). However, in a broad sense it also includes those born in the early 1950s.
13. Shakuto 2018.
14. Moore 2017.
15. Ono 2008.
16. Johnson 1982.
17. Yoshikawa 2021.
18. Yoshikawa 2021.
19. Dore 1973b.
20. Vogel 2020 [1963].
21. Dore 1978.
22. Smith 1978.
23. As documented, for example, in Takeda 2020, Klien 2019 and Obikwelu et al. 2017.
24. Kelly 1993a, 191.
25. Bestor 1989.
26. Hirayama and Ronald 2008.
27. Quinn and Turner 2020.
28. Hirayama and Ronald 2008.
29. Allison 2013.
30. Cook 2016.
31. Gill 2014.
32. Danely 2019; Kavedžija 2016.
33. Long et al. 2009.
34. OECD 2018.
35. Fire and Disaster Management Agency 2021.
36. De Togni 2021.
37. Littlejohn 2021.
38. Allison 2013.
39. Ministry of Internal Affairs and Communication 2020.
40. Kojima et al. 2017.
41. Ministry of Health, Labor and Welfare 2020a.
42. Hirai 2011.
43. Hirai 2011; Saino 1997.
44. *The Economist* 2019.
45. Miller et al. 2021.
46. Suzuki 2020.
47. Driessen 2018.
48. Higgs and Gilleard 2015.
49. Kavedžija 2016.
50. Lamb 2014.
51. Kavedžija 2019.
52. Lamb 2014.
53. Lamb 2014, 42.
54. Miller et al. 2021.

55. Okada 2005.
56. Murai 2014.
57. Byford 2012.
58. Miller et al. 2016.
59. Danesi 2017; Serafinelli 2018.
60. Miller 2015.
61. Pink 2001; Ginsburg 1995; MacDougall 1978.
62. Edwards 1992.
63. MacDougall 2005.
64. Lasén and Gómez-Cruz 2009.
65. Lindtner et al. 2011.
66. Ito 2005.
67. Ohashi et al. 2017.
68. Hochschild 1979.
69. Kavedžija 2018.
70. Malinowski 1923.
71. Goffman 1959.
72. Lim 2015.
73. Kavedžija 2018.
74. Roquet 2016.
75. As Brighenti 2007 argues, the relational, strategic and processual features of visibility invite exploration of it as a category of analysis in the social sciences.
76. Foucault 1977.
77. Lasén and Gómez-Cruz 2009; Bluteau 2021.
78. Ahmed 2010.
79. Kavedžija 2019.
80. Jackson 2012.
81. Hareven 2003.
82. Foreman 2017 [2008].
83. Brumann 2012.
84. Danely 2015a.
85. Ben-Ari 2013.
86. Kōchi Prefecture Tosa-Chō home page 2020.
87. Fujita et al. 2011.
88. Fujita et al. 2011.
89. Fujita et al. 2011.
90. Murakami et al. 2009.
91. Statistics Bureau of Japan 2015.
92. Statistics Bureau of Japan 2021.
93. National Institute of Population and Social Security Research 2016.
94. OECD 2020c.
95. OECD 2020c.
96. Mock 2012.
97. Traphagan 2004.
98. Tiefenbach and Kohlbacher 2013.
99. Tiefenbach and Kohlbacher 2013.
100. When explaining to people about my research and asking if they would like to participate, I gained oral consent and made it clear that they could withdraw from the project at any time. I listened to the wishes of my interlocutors about the stories they would like to remain private and have not included those in this book. When conducting formal interviews or when filming people, I gained written consent and provided them with an information sheet explaining the research in more detail.
101. I typically followed the wishes of my research participants in terms of the activities we did together. In Kyoto I found that my network rapidly increased from about half-way through fieldwork; by the end of the 16 months my diary was full of appointments, sometimes involving meetings with three or four different people in a day for walks, gallery visits, lunches, dinners and home visits. Conversations flowed naturally, ranging over many topics, and in this way I was able to get a sense of how care, ageing and digital technologies featured in their wider lives. I was able to accompany my research participants as they visited elderly

parents, attended temple events, met with friends and went shopping. The mundane activities on which I accompanied people were often when I learned the most about their daily lives. As I explain in more detail below, these relationships have continued online now that I have left the field; I regularly have video calls with several people. In total I came to know about 130 individuals, about 50 of whom I met regularly. It is impossible to count the number of conversations, experiences and moments that have contributed to this research. However, I can quantify the number of recorded semi-structured interviews with separate individuals (60). Of these, 25 were detailed explorations of participants' smartphones.

102. Collier 1957.
103. Gupta and Ferguson 1997.
104. I was fortunate that the ASSA project of which my research was part, provided me with funds for a research assistant. I benefited from having Sasaki Lise, a Japanese research assistant with a master's degree in Medical Anthropology, accompany me on all the trips to my rural fieldsite where being an 'outsider' was more conspicuous than in Kyoto. Previous ethnographic work on health in Japan has noted the tendency of research participants to 'perform' for outsiders. In Tosa-chō we conducted interviews together and people developed relationships to us both, so much so that a nickname for us emerged: Lau-Lise (*Ro-Rise*). Sasaki's presence facilitated my acceptance in this small rural community.

    Sasaki's role was primarily to assist with translation during interviews, and in their subsequent transcription. This was especially helpful in rural Kōchi where the local dialect, Kōchiben, is quite different from standard Japanese. I relied on Sasaki's translations as a bi-lingual Japanese person. However, when it came to particular terminology that became important in this book (such as the term '*yui*', meaning mutual support, or '*ikigai*', meaning purpose in life), it was through explanations from research participants that I developed an understanding of the meaning of such concepts in their lives.

    Sasaki emerged as more central to my rural research than can be encapsulated by the term 'translator'. The role of the research assistant in ethnographic fieldwork is under-theorised (Cons et al. 2020), with their contribution to authorship and what constitutes 'the field' often overlooked (Gupta 2014). In this research I learned how to interact, especially with older research participants, through observing and imitating Sasaki's manner. In addition, her convivial personality enabled rapport to be established quickly with people who may have been initially shy about talking to a foreign researcher. Together with Sasaki I have co-authored three publications (Sasaki et al. 2021; Haapio-Kirk et al. 2024; Haapio-Kirk and Sasaki forthcoming) and we continue to collaborate.

105. Through the adoption of graphic methods, my research aligns with the recent turn towards 'multimodal' research methods that have been described by anthropologists Dattatreyan and Marrero-Guillamón as 'multisensorial rather than text based, performative rather than representational, and inventive rather than descriptive' and, as such, can enable a 'more public, more collaborative, more political' anthropology (Dattatreyan and Marrero-Guillamón 2019, 221).
106. See for example Taylor et al. 2012; Hine 2020; Markham 2013; Murthy 2008; Sanjek and Tratner 2016; Favero and Theunissen 2018.
107. See for example Collins et al. 2017; Lapenta 2011; Pink and Hjorth 2012; Tacchi et al. 2012; Cruz and Ardèvol 2013; Murthy 2008; Postill and Pink 2012.
108. Pink 2012, 2011.
109. Favero and Theunissen 2018.
110. Pink 2012; Walton 2015; Favero 2018; Walton and Haapio-Kirk 2021.
111. Gubrium and Harper 2016.
112. I intended to document my fieldwork through drawing, but I quickly found that it was difficult to bring out a sketchbook in the kinds of intimate one-on-one social encounters through which much of my research was conducted. Unlike using my smartphone as a research tool which blended into the typical practices of everyone around me, I felt that holding a sketchbook, like writing in a notebook, marked me out as a researcher rather than a friend. I ended up limiting my drawing to where it was helpful – giving me an excuse to stand on street corners for extended periods of time while I sketched buildings in my neighbourhood and simultaneously observed the flows of people and traffic, or inscribing details I noticed during the health check, which may have gone unnoticed if recorded through photography.

    In this way, drawing turned out to be a good way of grounding myself in the field, but did not work as a regular method of inscribing research experiences as I had originally hoped.

However, I decided that if I felt unable to draw in social situations, perhaps my research participants could do so and we could use those drawings as the basis of elicitation exercises. The resulting drawings feature in Chapters 4 and 5, where I have extended discussions of them in relation to my wider findings. The other drawings I produced were given to my research participants as gifts, such as portraits of their homes or of their local mountain landscapes. The custom of gift-giving in Japan is strong, especially in traditional Kyoto and in my rural fieldsite, where I never seemed to come away empty-handed from visiting anyone. I therefore needed to be prepared with gifts, and handmade gifts such as paintings and drawings served well, in addition to a substantial quantity of bought gifts.

113. Hallam 2009, 83.
114. Banks and Zeitlyn 2015; Cox et al. 2016; Pink 2009; Stoller 2010.
115. Ingold 2007; Causey 2017; Taussig 2011.
116. Miller and Haapio-Kirk 2020.
117. Hogan 2017.
118. Takaragawa et al. 2019.
119. Literat 2013.
120. Hallam 2009, 77.
121. Hänninen et al. 2021.
122. Ardèvol and Gómez-Cruz 2013.
123. Hallam 2009.
124. Dix and Kaur 2019.
125. Literat 2013.
126. Literat 2013.
127. This approach is informed by Rattine-Flaherty and Singhal's (2009) discussion of the feminist values of visual participatory methods.
128. Pink 2015.
129. Stoller 2010.
130. Turkle 2011.
131. Kondo 1990.
132. Miller et al. 2016; Miller and Sinanan 2014.
133. Wang 2016.
134. As documented by Kavedžija (2019) in Japan, Garvey and Miller (2021) in Ireland and Walton (2021) in Italy.
135. United Nations 2019.
136. Hänninen et al. 2021.

Image by Laura Haapio-Kirk.

2
Experiences of ageing: beyond categories

> I'm not an elderly person … By socialising you don't have the time to think about things like that.
>
> Haruki-san, aged 85, Kyoto.

Haruki-san is 85. He is a tall man with a lively sense of humour who sees other people of his age around him becoming 'elderly', yet feels that his way of life allows him to avoid such categorisation. Every day he wakes up at 6.30 a.m. to attend prayers at his local temple in Kitano,[1] Kyoto. Haruki-san is a firm fixture at the temple and is treated with warmth and affection by the community there. He can often be found having lunch at the temple cafe with younger women, joining gatherings for his old Boy Scout network, in which they make hotpot and drink beer in the temple grounds, or attending his 'invention society' with his friends. He does not attend the local elderly society, however, as he feels that attending would, in his words, 'suck away my youth!' He explains that socialising, especially with young people, and staying curious is his secret for remaining healthy in later life. For Haruki-san, the category of 'elderly' is inherently separate from his own way of living. It is other people who 'become elderly' by staying at home, losing curiosity and thinking about death.

The streets of Kitano, busy with multiple generations, reflect the intention of many older people I met to remain visible and social outside of the home. They believe this to be a purposeful way of delaying negative aspects of ageing that they see as bound up with isolation. Anthropologist Iza Kavedžija argues that being a 'good older person' in Japan is a concept not far removed from Japanese social personhood more generally; at the heart of both lie a commitment to interconnectedness, self-cultivation and care.[2] In this chapter I show how contestation of the category of 'elderly' often revolves around social practices mediating intimacy and distance – many of which, with the rise of smartphones among older adults, are now moving online.

Haruki-san, to whom we will return below, is navigating later life in much the same way as he did in his younger years. He remains deliberately open to new experiences and expresses his curiosity about the world by staying active in his community. Independence and autonomy are often presented as cornerstones of 'successful ageing' – a concept critiqued as deriving from global processes of neoliberalisation.[3] A growing area of anthropological research seeks to find a middle ground among wider discourses framing ageing, either within the tropes of decline on the one hand or of 'successful ageing' on the other.[4] In Japan shifting societal attitudes towards old age stress self-reliance and responsibility,[5] resonating with this global focus on the individual as responsible for how they age. My research questions what independence for older adults really means in the digital age, given that people's lives and wellbeing are so intertwined with others across the life course.[6]

The modernisation of Japan, with its associated increased longevity and delay in the onset of the frailties of ageing, shifted the perception of when old age commences; it has also been argued to have reduced the status of older adults.[7] These shifts have occurred through the mechanisms of the dissolution of the extended family household, a devaluation of the knowledge and skills of older people and the increased burden of an ageing society placed on those of working age.[8] This chapter attempts to unpack these transformations – first by examining how categories of age emerge as culturally embedded phenomena, then by moving on to consider how individuals respond to such categorisation.

Through ethnographic examples of intergenerational interdependence, this chapter discusses the methods people employ to strike the right balance between autonomy and dependence[9] that my research revealed to be emerging in digital practices. I pay particular attention to the way in which relationships of care are shifting from the physical space of the home to digital spaces via the smartphone, developing an argument that will continue throughout this book in presenting the smartphone as a kind of domestic space in which we dwell. Following the introductory chapter, which outlined both the demographic shifts that have occurred in the Japanese population as it has become a 'super-ageing' society and the public discourse surrounding those trends, the discussion turns to the experiences of people themselves. I aim to bring to life the stories of the individuals that make up the ageing society of Japan, challenging the imposition of age-based categories by showing how people both resist and embrace them, increasingly through the adoption and rejection of digital technologies.

Categories of age

In post-war Japan, when many of the people who participated in this research were born, life expectancy was relatively low. In 1947 the life expectancy for men was 50 and women 54,[10] whereas in the UK it was 64 and 68 respectively.[11] As recently as the 1960s life expectancy in Japan was the shortest among the G7 countries. Yet after a period of rapid economic growth in the latter half of the twentieth century, Japanese life expectancy underwent dramatic lengthening; its population now regularly ranks among the world's longest living.[12] In 2018 the average life expectancy was 87 years for women and 81 years for men.[13] A rapidly extended lifespan presents new opportunities for people in mid-to-later-life, with many pursuing lifelong learning,[14] and continued employment,[15] alongside personal hobbies and interests.[16] Yet people also have to contend with increasing care demands as they look after older generations, often while having to face the advent of frailty and the inevitability of death. Such a trajectory is multi-faceted and takes many forms.[17]

In 2017 the official classification of old age in Japan was adjusted from 65+ to 75+ to reflect the prolonged period of health and vigour that much of the population now experience after retirement.[18] Based on research on the physical and psychological health of those in later life, the Joint Committee of the Japan Gerontological Society and the Japan Geriatrics Society suggested that those aged between 65 and 74 years old should be re-classified as 'pre-old age' (*zenki kōrei-sha* or 'early-stage elderly') and those aged 75 and over as 'old age' (*kōki kōrei-sha* or 'late-stage elderly'). People aged over 90 would then become known as 'super-elderly' (*chō kōrei-sha*).[19]

While age-based terms such as *kōrei-sha* ('elderly'), *shinia* ('senior') and *siruba* ('silver') can be useful for developing services and health and care policy, I found that such categories are of diminished importance for most people in their lives (for a discussion of these and other Japanese age-based terms, see note[20] and note[21]). For example, people may recognise that they are suffering from age-based illnesses or use the category of 'elderly' when struggling with their smartphone ('I can't do it, I'm elderly!'), yet in their day-to-day lives they also enjoy many continuities with their younger selves; for many, interests and personality seem to have become even more pronounced with age. However, one of the areas in which people did feel the consequences of belonging to age-based categories was in employment, especially as they near the mandatory retirement age of 65. I discuss this concern in Chapter 3, which focuses on work. The proposal to shift the official categories of old age was intended to stimulate policy reform around the productive economic capacity of the 'pre-old', and thus to reduce the socio-economic burden of an ageing society on younger generations.[22]

The emergence of 'pre-old' as a proposed policy category in Japan aligns with the wider academic concept of the 'third age',[23] generally considered to be the period of life between retirement and a 'social imaginary' of a so-called 'fourth age' that signifies the beginning of age-imposed limitations and frailties.[24] The boundary between the two remains undefined; indeed, as social psychologists Paul Higgs and Chris Gilleard note, in affluent post-industrial societies the 'fourth age' exists more as a shadowy opposite of the 'active' and 'successful' third age rather than as a distinct set of criteria.[25] Such academic discourse on age categorisation reflects Global North social imaginaries of ageing, in which vulnerability and frailty are often posed in opposition to productivity and robustness. 'Real' old age as a period of life is thus pushed to an anticipated time in the future, marked by loss of agency and decline.[26]

The othering of ageing bodies that condenses 'real' old age into a life stage that is always on the horizon, inherently separate to prior modes of being,[27] is not dissimilar to the way in which disabled people experience othering, both in Japan more broadly. Sociologists van der Horst and Vickerstaff ask the pertinent question, 'is part of ageism actually ableism?', suggesting that these intersecting concepts should be analytically separated in order to have a clearer understanding of both.[28] They argue that the concept of the 'fourth age' contains hidden ableism that must be unpacked. Many older adults of course do experience age-related disabilities and health conditions that intersect with their experience of ageing; however, it is important that underlying ableist attitudes are addressed in societal discourses on ageing.

Ableist narratives were underneath many of the distinctions people made between themselves and 'other people who age', as reflected in Haruki-san's negative feelings towards people who 'stay at home', presented above. However, there were also benefits to ageing that Haruki-san alluded to. He did not perceive his oppositional viewpoints as being contradictory.

As argued by Verbruggen, Howell and Simmons, it is up to anthropologists to co-create narratives of ageing with research participants that go beyond binaries and show the 'in between'.[29] Physical and mental decline certainly concerned many of my research participants who could be considered to fit the criteria of 'pre-old age' or a 'third age'. However, they voiced more pressing concerns associated with anticipated longevity, such as financial insecurity, and a desire to stay away from hospitals where 'they will not let you die', as expressed by one woman in her sixties from Kyoto. Here the concern was less about her own personal decline, rather about the surrounding institutional structures and societal attitudes towards the end of life that she felt might impede on her sense of agency. The distinction between 'third' and 'fourth' ages serves to demarcate a 'normative life course' based on periods of productivity.[30] Emerging in the 1960s, a 'life course approach' was adopted by demographers, sociologists and psychologists as a way of understanding how people's lives follow certain patterns and progressions; a 'life course' can be defined as 'a sequence of socially defined events and roles that the individual enacts over time'.[31] Over subsequent decades this study of the life course has been taken up, critiqued and expanded by anthropologists,[32] who have moved from a gerontological focus to revealing the complexity of ageing across generations, showing how 'transitions of aging are a matter of entangled horizons of transformation'.[33] David Plath pioneered the anthropological study of the later life course in Japan, revealing how middle-aged individuals were already adapting to increasing longevity and the '80-year life' in the 1970s.[34]

Recent research adopting a life course approach in Japan recognises the importance of maintaining a fluid approach to the categorisation of life stages, seeing them as diverse, individualised and subject to change.[35] The anthropologist Jason Danely has demonstrated how in Japan the later life course is situated in relation to Buddhist notions of lifecycle that go beyond death.[36] In Chapter 8 I will return to the concept of progression through the life course in relation to life purpose (*ikigai*), arguing that the two have become conflated and showing how key moments in life can serve as 'turning points' towards meaning.[37] I met many people in their sixties and seventies whose life course did not match normative models, either by remaining single, getting divorced, starting new careers or entering new relationships later in life. These examples impact our understanding of the life course itself. Rather than view the life course as a series of logical progressions, it has become less predictable and more diverse, compounded by changes in the wider context of a political economy that contributes to a general sense of unpredictability – including the bursting of the 'bubble economy' and the precariousness of the current labour force.

I found that people in their sixties and seventies, after spending their lives following more normative models of life course progression through family and work life, are often increasingly influenced by the more fluid life courses of their children's generation. The latter frequently challenge norms of what is expected as one ages by delaying entry into full-time work and marriage, and thus entry into societal notions of adulthood, sometimes indefinitely.[38] The resultant dialogue between the generations, sometimes taking inspiration from each other, but also sometimes acting in deliberate contrast to each other, informs how people of all generations view their life course and imagine their futures.

One woman in her forties suggested that her parents' generation had it much easier than she does, precisely because they had less choice in determining their life course; there were only a narrow range of options for baby boomers, expected to fulfil the ideal nuclear family model. By contrast, she felt that her single life, in which she was striving towards a career as an artist and academic, was more troubling and uncertain. She could not help but view her lack of progression through expected life course events such as marriage as a kind of failure, even though she felt that this had arisen from her decision to follow her own desires rather than conforming to the expectations of her parents and society.

Discourses of ageing and generations

As anthropologists Pina-Cabral and Theodossopoulos note, the very notion of 'generation' is to be critiqued for its lack of nuance. However, the concept has been of value in social analysis since the early twentieth century.[39] Mannheim, an influential German sociologist from the first half of the twentieth century, describes the formation of generations through shared social and historical circumstances, giving rise to a shared generational consciousness.[40] However, sociologists and anthropologists have documented the various ways in which individuals within a given generation may feel socially distant from one another.[41] In Japan, generational stereotyping has been part of popular discourse since at least the *Meiji* era (between 1868 to 1912), gaining momentum after the Second World War.[42] For example, during the post-war period generational discourse focused on responsibility for the war, with the people born in the early *Showa* era (between 1926 to 1989) emerging in the popular consciousness as the bedrock of post-war recovery.

The anthropologist William Kelly argues that this generation became a standard against which later generations were judged. Among these were the baby boomers, also known as 'new family types', who gained a reputation for being concerned only with the personal lifestyle advancements of the high growth decades and lacking experience of the hardships endured by previous generations.[43] Popular discourse around the baby boomer generation was amplified in 1976 with the publication of a novel by Japanese author Sakaiya titled *Dankai no sedai* ('generation of a mass'). It focused upon the cohort born between 1947 and 1949,[44] but the terms *'dankai no sedai'* or *'dankai'* have subsequently become used more broadly in Japan to indicate the post-war generation without a clear-cut definition of the term.[45] Only in the 1990s did the term start to gain wide prominence in public discourse in Japan in anticipation of their mass retirement.[46]

Media portrayals of baby boomers over the past 20 years in Japan have varied. The cohort has posed a societal challenge for welfare provision while simultaneously representing an opportunity for continued work and participation in volunteering.[47] The Japanese media tend to praise active baby boomers, regarding them as a social resource, and noting in particular the emerging advantage of women in later life compared with men.[48] At the other end of the spectrum are articles and television programmes about the 'no-connection society' (*muen shakai*), reflected in such phenomena as elderly people dying alone (*kodokushi*); these portrayals present a disturbing picture of a society whose traditions of family and respect for the elderly are considered to be diminished.[49] The post-war cohort of 'baby boomers' has been the subject of rising global critique. In 2019 the phrase 'OK Boomer' surged in global pop culture, with Millennials (born between 1981 and 1996) and Generation Z (born between 1997 and 2012) using the phrase in digital spaces to channel a wider frustration with underlying socio-economic and ecological problems.[50]

The Covid-19 pandemic has also served to catalyse generational tensions online in Japan. A story exploded on Twitter in July 2020 that captured the Japanese media's attention. According to the original tweet, a woman in a supermarket heard an older man reprimanding a young mother for buying ready-made potato salad, asking 'If you're a mother, why don't you cook (something as easy as) a potato salad yourself?' The mother stood there with her head down, still holding the salad, at which point the Twitter user picked up two packs of the same salad to try to communicate to the woman that it was all right to buy it. The tweet went viral, and for days the hashtags #potatosalad (*#potetosarada*) and #PotatoOldman (*#PoteJii*) were trending on Japanese Twitter, with others sharing similar experiences.

The potato salad episode illustrates the intergenerational tensions that are present in Japan. These can emerge in everyday interactions, but are more directly revealed in online and other media discourses. The elderly man in the supermarket went unchallenged in the interaction, yet anonymous Twitter accounts provided a space in which (younger) people could vent their feelings freely. These spaces have proven highly active during the Covid-19 pandemic, such as in online tabloid stories about 'selfish' older people hoarding masks.[51] Criticism was aimed at seniors who queued early in the morning at pharmacies to procure masks, with the accusation that this behaviour made it more difficult for people of working age to obtain them – despite the availability of masks online, which may have privileged younger shoppers.

As anthropologists Takahashi Erika and Jason Danely point out, in these media discourses we witness how crises, such as the Covid-19 pandemic, can naturalise moral judgements about older people,[52] as well as contribute to a widening division between generations. Indeed, anthropological research has argued that the very idea of generation comes to the fore most strongly in moments of upheaval and social change.[53] However, my research shows how older adults are starting to occupy these online spaces, putting them in a better position to nuance generational stereotypes and level communication between generations.

Life beyond categories

When I began this research in 2018, the Japanese government had just established a council for designing the '100-year life society' (jinsei 100 nen jidai), focusing on economics and human resources.[54] As Chapter 3 will explore in relation to the continuation of work beyond retirement, the government's response to mass longevity in many ways parallels the concern of older adults to remain in work to stay healthy. Yet simultaneously it draws attention to the increasing financial precarity facing both the nation and individuals as they age.[55] Living a very long time was generally viewed with apprehension by many of my research participants because of the fear of physical and mental decline – an anxiety made more acute by the wider situation of uncertain pensions and healthcare costs facing those nearing retirement age. However, I found that for many people the sense of decline that may correspond with the categorisation of 'elderly' often aligns less with chronological age than with a gradual shift in attitude. There was a common belief that no matter how many years old you are, if you no longer have interests and passions in life you will age. One must have purpose in life (ikigai) to remain healthy – a conviction that will be clarified and expanded on in the penultimate chapter of this book.

Haruki-san, who was introduced at the beginning of this chapter, speaks his mind and experiences continuity with his younger self, whom he regards as having been honest and open. Yet he also recognises that ageing comes with a certain letting go: he does not care so much what people think of his meetings with younger women, or of him being loud and drinking beer. As he observes

> You shed your skin … you become a better person as you age, and you don't care about others.

Yet in fact Haruki-san does care about others. He is a leading member of the Boy Scouts network run from the temple, even though his sons, once members themselves, are now adults and have moved away. He is a key organiser of social events and will message people on LINE to make sure they know the time and place of the get-together or cherry blossom party (*ohanami*) that he is organising. The kind of care that Haruki-san no longer participates in is caring how others may judge his provocative nature. However, at the same time he practises care by maintaining his responsibilities to his various communities and to his family business.

Like many other older people in this research who continue some form of business or employment after retirement, Haruki-san maintains an active role in the family fabric business, aided by his smartphone, which he bought on a deal three years ago. His LINE messaging account is full of photographs of the fabric that he sells online in partnership with his son; the latter uploads the product photographs to an online marketplace while Haruki-san receives the sale requests on LINE. Every day he makes three or four sales. Through his smartphone, he sends customers photographs of rolls of fabric with stickers attached to them to indicate the grade of the material; he then completes the orders by posting the fabric. Haruki-san's sons set up this system for him, and although he still protests that he does not know how to use his smartphone and prefers to use his old feature phone (*garakei*) to make calls, he actually uses his smartphone very efficiently every day for taking photographs and accessing the internet.[56] Haruki-san's participation in social, family and business life is often interwoven with his smartphone use. It seems as though the device is facilitating the kind of activities that he claims will save him from becoming elderly, thus enabling him to remain 'pre-old' perhaps indefinitely.

As an active and social 85-year-old, Haruki-san could be seen as a model of 'successful ageing'. He is doing everything that is officially advised for older people to remain healthy, according to the Kyoto city website which offers health and welfare guidance for different age categories.[57] Rather than viewing his activities as a form of 'successful ageing', however, he sees them as merely continuities with his youth and middle age. This denial of 'elderly' status may be prompted by the wider societal negative appraisal of ageing. On the city website, for instance, reaching the age of 65 was described in terms of crossing a threshold at which one suddenly becomes elderly (*kōrei-sha*), denoting a period of life when people might become 'anxious and lonely' and might therefore require advice on how to remain active in society and find purpose in life (*ikigai*). Yet the kinds of official activities set up by the municipality to promote sociality among older people are precisely the activities Haruki-san avoids. He feels that by attending he would join a group of people who have acquiesced in being categorised as 'elderly', which for him would represent the start of age-based decline.

In addition, the official discourse appears to be 20 years out of step in terms of Haruki-san's life course. For example, illustrations of grey-haired characters engaged in hobbies that adorn the city website bear little resemblance to the busy lives of the 65-year-olds who participated in this research. Having purpose through roles in business and volunteering is central to how Haruki-san maintains his sense of self as he approaches his nineties. Yet he does not describe these activities explicitly as his '*ikigai*', as Chapter 8 will show is the case for many people.

Haruki-san's continued participation in the family business depended on the actions of his sons to enable his digital communication with customers. Yet this initial dependency has disappeared as his confidence in using the smartphone grows and he uses LINE to communicate with people more broadly in his social sphere. By establishing a daily schedule, Haruki-san has re-created a working day, with responsibilities and sociality beyond the home. It is this schedule which confirms to Haruki-san that he is not elderly – a status to which he may only admit if and when his physical state declines and he becomes housebound. For him, staying at home is what elderly people do.

The home thus can become tinged with negative associations. As I will show in the later examples in this chapter, the home can be a contentious place, particularly for men after retirement, when married couples have to adjust to sharing domestic space often after a lifetime spent apart. Yet Haruki-san also illustrates a new possibility: that the smartphone would allow him to maintain a sense of agency and social participation even if he was unable to continue his activities outside the home.

The next section presents further evidence on the ambiguity of the home for retired men. For them it represents a space of social withdrawal, ultimately prompting a drive towards continued employment or volunteering, yet simultaneously being a space of care. I then show

how parallel processes of negotiating gendered dependency and retired identities are reflected in smartphone usage among men and women. As such, the smartphone can be considered as an extension of the domestic sphere, where family relationships and emotional labour are still largely the responsibility of women. As the next section illustrates, the historical ideal of multigenerational living presents its own challenges of over-closeness – something that I found technologies to be increasingly mediating.

Multigenerational households

The Japanese word *'ie'* refers to the traditional ideal of the patriarchal family, consisting of grandparents, their eldest son and his wife, and their grandchildren.[58] The word, which also means 'household', refers to the physical estate, such as a house and associated land, where the family reside. The term still has considerable significance today, even with changing living arrangements, since the family registration system, based on a household with a head of the family, revolves around this traditional principle. Along with inheriting property, the eldest son (or more accurately his wife) also inherits the responsibility to care for ageing parents. Older women tended to agree that multigenerational living can be difficult, especially if relations between mothers and daughters-in-law are strained. As one woman in her sixties from Kyoto who was caring for her mother-in-law put it: 'Being the wife of a first-born son is very difficult'.

For previous generations, such a position in a traditional, multigenerational household meant that a woman was effectively the 'wife of the house',[59] freeing the older mother-in-law of the responsibility of cooking and cleaning for the household and in some cases 'becoming like a slave', as another woman put it. Several female participants in their fifties and sixties expressed regret for the way their mothers were treated by the families they moved into upon marriage, often moving far away from home to find themselves at the bottom of a household hierarchy. There was a sense that as these women got older and experienced increasing care demands themselves, they understood the struggles their mothers had faced more fully. Others had deliberately remained single to avoid such experiences entirely.

Influenced by the philosophy of Confucianism originating in China, like many neighbouring East Asian countries, Japan has deeply rooted traditions of filial obligation involving intergenerational co-residence,[60] though these have declined in recent years with the advent of Long Term

Care Insurance.[61] The rapid extension of later life in Japan during recent decades, taking place over the duration of just one lifetime, has meant that the expectations of filial piety, while previously relatively short-term, now are elongated with an unknowable duration. Social and familial expectations around care have been subject to rupture and instability on a personal and national level. Anthropologist John Traphagan, drawing on extended fieldwork in rural northern Japan, argues that discourses on filial piety are not fixed; they are rather continually constructed and reconstructed by children and their parents in the context of eldercare.[62] Demographic shifts, changing attitudes towards co-residence and the increasing participation of women in the labour force have meant that in recent years the responsibility for care of the elderly has become fraught with contradictory sentiments as people try to balance their continuing loyalties to family with a desire for privacy and autonomy. This autonomy is not only desired by middle-aged people but also by those who are older and in need of increasing care, yet with a strong desire to avoid becoming a burden on those around them.[63]

A sense of turbulence regarding eldercare has been reflected in the way in which the government has gone back and forth on the balance between state and family care provision for the elderly. Changes to the Civil Code in 1948 reflected a move away from the codification of Confucian values of filial piety in law. However, this change was brought about by the American authorities who occupied Japan at the time, and thus did not reflect the values and practices that remained in Japanese families.[64] In addition, even with these changes in law, children were still legally responsible for the care of their elderly parents in the new Civil Code.

In 1987 the government released a White Paper on Health and Welfare aimed at mitigating the anticipated growth in public medical expenditure associated with an ageing society. In this document the government recommended that the responsibility of caring for elderly patients should be shifted from hospitals to the family.[65] The 'Golden Plan', launched in 1990 to assist with this endeavour, provided in-home care for the elderly, assisting family caregivers – typically middle-aged women – given the increasing participation of women in the labour force. Yet in 2000 the government launched the Long Term Care Insurance scheme, which aimed to unburden the family of care for the elderly and place more responsibility of care on the state. However today, with increasing demand placed on an overstretched system, many people speak about the long waits associated with accessing care and the process of needs-based assessment that can exclude people deemed not to be in severe need of care. In practice, I found that the responsibility of care for the elderly falls primarily on middle-aged and older relatives, generally daughters and daughters-in-law.

The practice of multigenerational co-residence was the mechanism through which family care was traditionally managed; it was the norm for half the population as recently as 1995.[66] The term *nisedaijūtaku*, literally 'two-generation housing', was coined by the housing company Asahi Kasei Homes in 1975 when they launched a new line of prefabricated housing designed for two semi-separate households in one building.[67] The concept was promoted in the context of the post-war rise of the nuclear family and the associated housing boom that saw land prices increase; it made financial sense to many to take advantage of building on their parents' land. By allowing people to maintain independent households yet provide a supportive presence to each other, the idea of two-generation homes caught on and became a significant proportion of the housing stock. As we shall see in the ethnographic example to follow, people develop ingenious ways to manage their privacy and connection within multigenerational homes, as I observed during my stay in one such household in Osaka.

Not *mazacon*!

The Katsuras – Takashi-san and Keiko-san – are a married couple in their mid-sixties who live in a suburb in the north of Osaka city. As noted in the previous chapter, I stayed with them for two months when I first arrived in Japan. During the time when I lived in the house, the couple both worked in hospital administration; they left home early in the morning and commuted to work by train. Their house, a three-storey detached building overlooking the tracks of a major train line, was built 20 years ago on Takashi-san's parents' land when they knocked down the previous building, as is customary to do in Japan periodically.[68] While from the outside it looked like simply a large house, the building was internally divided into a two-floor maisonette on the top two floors, where the middle-aged couple lived, and a self-contained ground floor apartment, where Takashi-san's parents lived. The households were connected by a sliding door situated between two separate entrances (Fig. 2.1).

With the building containing duplicate kitchens, toilets and bathrooms, there was limited interaction between the two households daily. The older Katsuras, who lived downstairs, were in their mid-to-late nineties and during my fieldwork were highly independent. The older Mrs Katsura went regularly to the nearby supermarket alone, pulling her grocery trolley behind her; she also cooked all the meals for herself and

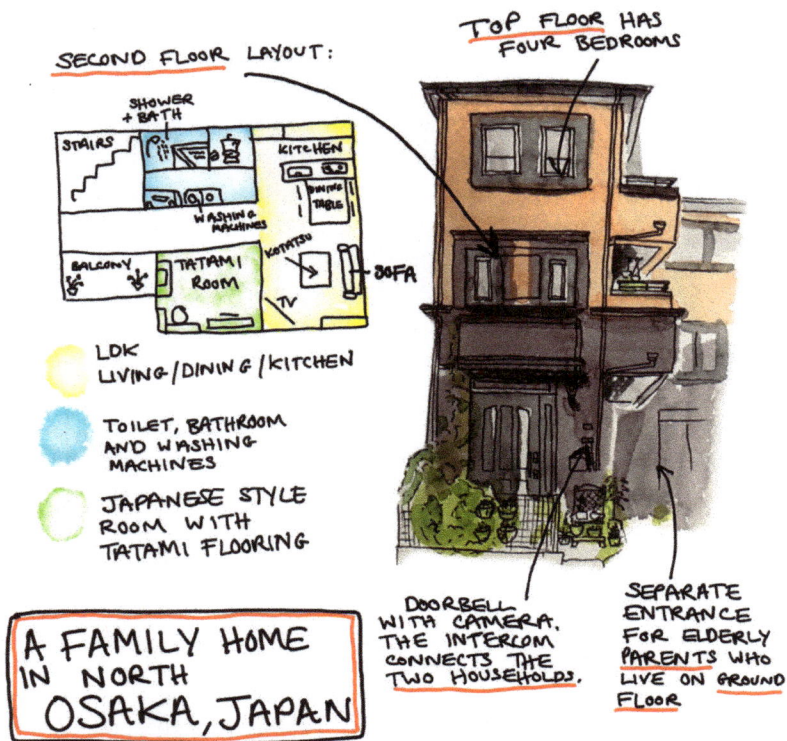

Figure 2.1 A sketch of the Katsura residence. Image by Laura Haapio-Kirk.

her husband, who attended the nearby temple where he was an active member of the community. The senior Katsura-san was a skilled artist. At the age of 80 he painted a series of large, elaborately decorative murals featuring fauna and flora which were installed in the ceiling of the temple. Now, nearly 20 years later, he rarely painted at all due to failing eyesight. However, when I gave him a watercolour painting I made of the house at the end of my two-month stay, the next day he made me a painting of plum blossom in return.

While there was limited face-to-face interaction between the households, the two generations of the Katsura residence regularly communicated via an intercom system attached to the wall in the LDK (Living-dining room-kitchen, see house layout shown in Fig. 2.1). This system was connected to the front door. A camera doorbell allowed the family to see who was at the door, while also enabling the two households to communicate without having to go up and down the stairs. The intercom was used in place of a mobile or landline phone since the older Katsuras did not own a mobile phone and the intercom was conveniently positioned in the main living space of the home.

In this manner, technology was employed to facilitate independent living while simultaneously used for mutual caregiving. Care at a distance could therefore be witnessed even within one household. In that sense, the smartphone can be considered to work in continuity with previous and parallel technologies such as the intercom.[69] In the evenings the intercom would sometimes buzz and Taskashi-san would typically answer. It was usually his mother informing him of her day's activities, and often letting him know that she was going to bed.

On the evening of *Setsubun*, a festival celebrating the arrival of spring in February, we were eating dinner and the intercom was buzzing more than usual. With every update from Takashi-san's mother, his replies of '*hai!*' ('yes') grew increasingly deflated as he repeatedly got up from the dining table to answer. Sitting down again after the fourth or fifth buzz, he exclaimed half-jokingly 'not *mazacon!*', using the Japanese-English contraction for mother complex. This term, referring to the intense relationship of interdependency that could develop between a son (often the eldest) and his mother in middle-class post-war Japanese households, had become a cliche by the 1980s. Laughing, Takashi-san repeated the phrase after his mother suddenly appeared on the landing, the only time that I saw her upstairs, bearing festival food from the local temple.

In jokingly declaring 'not *mazacon*', Takashi-san was denouncing a stereotyped mother-son relationship and asserting both his and his mother's independence from one another. Yet at the same time he also acknowledged the sometimes intense nature of intergenerational kin relationships. The physically divided household was a material manifestation of the desire to balance privacy and intimacy held by both generations. In day-to-day communication, the intercom was the technological means through which this balance was most effectively achieved.

Over the subsequent 16 months of fieldwork it became apparent that this episode reflected something broader than the family dynamics of one particular household. It encapsulated the way in which performing care and mitigating burden are concerns intertwined across generations and increasingly played out via technology. Anthropologist Inge Daniels has demonstrated how being 'separate together' is crucial to Japanese intimate sociality within the home.[70] A core argument of this chapter is that, like the intercom in the Katsura household, the smartphone is becoming integral to the ways in which people express care while maintaining autonomy and minimising the burden of domestic duties, which now stretch beyond the physical home.

Takashi-san and Keiko-san feel that they are among the last generation who will practise multigenerational living in the sense of sharing a physical home. Such a conviction is a common refrain from my participants in their sixties and seventies. People mostly explain this view as one of not wanting to be a burden on their children in old age, and a sense of not wanting to pass on the weight of tradition that they may have experienced negatively. However, several participants currently in their sixties who had children living elsewhere expressed the view that while this living arrangement might be fine for now, they would hope that in the future one of their children would return to the house. This move was desired not only to help them towards the end of the lives, but also to take on responsibility for the house itself. 'If we are not well, I hope someone will go home to protect our house (*ie o mamotte*),' explained one man in his mid-sixties, whose modern home had been built on his grandmother's land in central Kyoto. Care thus extends not only towards people but also to the material manifestations of lineage and ancestry represented by family land.

After returning from fieldwork, I remained in contact with the Katsura family. During the early stages of the Covid-19 pandemic in March 2020, Keiko-san informed me that her father-in-law had been admitted to hospital with a broken leg. He was 98 at the time and ended up staying in hospital for five months for rehabilitation. Because of the pandemic, he was not allowed any visitors. The family were worried that he would mentally deteriorate by not being at home interacting with them. Keiko-san felt that his isolation was compounded by the fact that he did not have a smartphone. She would bring him fresh underwear every day to the hospital and ask about his condition when she handed it over to the nurse. This difficult period was a worrying time in which remote family care was curtailed because of a lack of technology.

After five months Katsura-san had recovered enough to go home, just in time for his auspicious 99th birthday celebration (*hakuju*). 'We are so happy to have him home,' said Keiko-san. She was impressed at her father-in-law's ability to recover from such a disruptive and potentially isolating time, attributing his fortitude to his Buddhist daily practice of reading and reciting sutras which he had maintained while in hospital. Even during his hospitalisation, Katsura-san did not neglect his care for the generations above him through his Buddhist religious practices centred on ancestor worship. It was this ritualised act of intergenerational care between the living and the dead that may have inadvertently provided a form of self-care. Now that Katsura-san was experiencing more frailty, his own agency, expressed through acts of care, became even more important.

When Katsura-san returned home, the family kept a close eye on him and provided help if needed. His doctor remarked that it was astonishing that after a five-month stay in hospital he did not need to enter a care home but could still live at home. His mobility was restricted, but he could still move about the home independently, watch television and read newspapers. Posing for a photograph taken with Keiko-san's smartphone (Fig. 2.2) allowed others to share in his birthday celebration

10 likes

Figure 2.2 An Instagram post celebrating the birthday of Katsura-san. Screenshot shared with permission of anonymous research participant.

remotely. This photo, subsequently edited and posted on Instagram by Keiko-san, was taken while the pandemic was still ongoing and people were encouraged by the government to show 'self-restraint' (*jishuku*) by limiting movement and staying at home.

Despite this isolation, by posting the photograph on Instagram Keiko-san was able visually to demonstrate how her father-in-law was embedded in relationships of care. In this way, digital visual content has the capacity to shift the 'social imaginary' of the fourth age, making it less shadowy and showing how people continue to experience agency and have active social lives even as they experience increased dependency. While Katsura-san may not have had a smartphone, the primary reason why many of my interlocutors were using one was to stay connected and maintain their status as social persons, something that became particularly pertinent as they got older.

Both the intergenerational living arrangement, Katsura-san's involvement with the temple and his own religious activities meant that he was sufficiently socially embedded as not to require digital connection. Yet for the younger Katsuras, who are not involved in such activities, and for whom the smartphone is a central means of communication with their children and friends, the device is set to become an important part of the way in which their own experiences of ageing and care are mediated, as will be demonstrated in the next section.

Digital multigenerational living

Some months after Katsura-san's return home from hospital, he sadly passed away. Keiko-san explained that they were concerned that his wife might fall into decline after his passing, so she was taking extra care to help her stay healthy. When Keiko-san wakes up at 6.00 a.m. every morning, the first thing she does is to make a smoothie for herself, her husband and now also for her mother-in-law, which she takes downstairs to her. The morning smoothie was part of a regime of care, with the ingredients of apples, bananas, Japanese mustard spinach (*komatsuna*) and soy milk all chosen because Keiko-san believes them to be health-giving ingredients. As she explained:

> I think it's very good for our health. I want my mother-in-law to enjoy the rest of her life happily and calmly.

For Keiko-san, food was at the centre of health and of how she practises daily acts of care for herself and her family. The same was true for many of the people I met. Following Keiko-san's husband's stomach cancer 17 years ago, she paid particularly close attention to the food they ate and bought him special, gluten-free snacks. After dinner she would often pour puffed corn into a bowl for him to enjoy while watching television.

The association of food with care is now migrating online. Keiko-san's adult son often sends her photographs of the meals he is eating. Sending meal photographs between family members was a practice I observed with multiple participants, especially when they did not live together. In sharing the photographs of food, there is a sense of connection with the domestic practices of cooking and dining that occur in the home, which in Japan are mostly associated with the kind of care performed by a mother. Sharing meals is central to family life among the people I was with, whether at home, in restaurants or remotely through digital communication.

One photograph Keiko-san received showed her grandson behind a box of oysters that his father's friend had posted to them. Posting food is a common practice in Japan; I observed all kinds of local fresh food being posted across the country, often between relatives, from large boxes of citrus fruit to mountain herbs to sake. Keiko-san was unable to share physically in the consumption of the oysters, yet through her grandson's enthusiastic expression in the photograph she could join in the delight of receiving them. Keiko-san's husband also uses LINE to communicate with his sons, but not so much as Keiko-san. She felt it was better when they had a family chat group because then her husband could see all the messages too. However, the family chat group has fallen out of use and now she messages her two sons independently.

Anthropologist Inge Daniels, drawing on her ethnography within Japanese homes, argues that the production of a home-like atmosphere rests on moments of heat and proximity between family members, such as warm baths and huddling together under a *kotatsu* heated table.[71] However for Keiko-san and many of the people I got to know during my research, her sons and grandchildren visited her home occasionally; it was through daily interaction in digital spaces that intimacy and a 'homely' atmosphere was actively maintained. In Chapter 5, which considers visual digital communication, I will elaborate on the mechanisms that generate this digital affective atmosphere. It was Keiko-san's digital affective labour that maintained a sense of proximity and closeness between family members. As we shall see in the next example, involving a couple in Kyoto of a similar age, there are various possible reasons behind the gender-divided usage of the smartphone by older Japanese adults.

Saori-san and her husband Akira-san are a couple in their mid-sixties from Kyoto. Following retirement, they spend their time managing their property business, travelling and enjoying their hobbies. Their two daughters live abroad, so they visit them for several weeks at a stretch. In the meantime they stay connected through LINE. As Saori-san observed:

> Yeah, LINE we use the most! We can send photographs on LINE too and call. Everything is on LINE.

Their tennis group was also co-ordinated through LINE, as was Akira-san's golf group. However, instead of them each having their own individual LINE accounts, as was the case with most of the people I met who had a smartphone, Akira-san did not have an account; everything was centralised on Saori-san's account. She has an iPhone and they bought an iPad for Akira-san and installed LINE on it, though he never engages in messaging. Saori-san mainly used the iPad a lot when at home; the large screen size was ideal for looking at photographs of their grandchildren, and she felt it was much easier to type on than a small smartphone keyboard.

As we were talking about LINE with Saori-san, her husband chimed in, declaring 'I can't relate to any of this conversation! I only call! On my *garakei*!' Akira-san had a simple feature phone (*garakei*) that allowed him to call and reply to text messages, though he would never initiate a text. When his golf group sent a message to Saori's LINE account, he read it and asked her to reply for him. In the conversation between husband and wife transcribed below, Saori-san and Akira-san give their perspectives on the reasons behind their particular digital practices.

> Saori-san (to her husband): You don't try to understand the convenience of it [the smartphone]. So, when your *garakei* broke, you changed it to another *garakei*! Our elderly friends were like, 'Why?!' So there are many types of people, those who want to know more about new things and new conveniences and want to be caught up with the trend. And then there are others who don't care, and are happy with the way it is now, and can't be bothered to learn new ideas.
>
> Akira-san: Yes very true! I like that! … The people I play golf with are a little younger. And for those people, they need to be included in the LINE golf group so they need to have a smartphone. They ask me to change to a smartphone! But it's not like I'm going

to be a part of their group. When I used to work there were people I would go out with. But it's not like that any more. And I just happen not to have friends who are on LINE.

When Akira-san was working he used a pager (*poketto beru*) and his *garakei*. He claimed that he would not have been able to do his job without those communication technologies. Now that the technology has moved on, he is retired and lacks the social and practical context of work to necessitate a shift to the smartphone. His friendships that he developed during his working life over the past 40 years were all with other men like him. They were similarly reluctant to get a smartphone, so there was no social reason to make the switch – especially when his wife completed any necessary digital sociality on his behalf. Saori-san explained her husband's behaviour as being down to a lack of interest or desire to learn, yet Akira-san's reasoning concerning work and his social circles resonates with the experiences of many of my older male participants in this research. Similarly reluctant to switch to a smartphone, they felt no social pressure to participate in digital communication. The gendered dimension to smartphone usage is not surprising, given the stark divides that separate men and women in many areas of life in Japan – from employment[72] to social life[73] to the very way the citizens are recorded through the *koseki* family register system.[74]

Anthropological work has shown how in Japan the social expectation for women to be nurturers and mother-figures for their husbands can be damaging not only for women but also for the men who perform dependency (*amai*),[75] linking with the discussion of *mazacon* above as a contested phenomenon. In many ways, the dependency that Akira-san has on Saori-san for managing his online social interactions via her LINE account represents a continuity with a dependence of husbands on their wives in the domestic sphere, shown in activities such as shopping and cooking. Some men performed an almost child-like dependency on their wife as the mother-figure, for example by holding up an empty mug and saying 'Mother, coffee!'

In retirement, the domestic space can become negatively charged for men, as we saw with Haruki-san who equated staying at home with a swift demise (p.39). Male dependency on women as the head of the home may have been necessary for men during their working lives, when they spent long hours outside of the home, but it can become intolerable to some in retirement. I argue that the rejection of the smartphone by older men such as Akira-san is driven by an unfamiliarity and, in some cases, a fear of domesticity – but it also continues this expression of dependency. The smartphone represents a hub for sociality among kin, much like the traditional multigenerational home; in this new digital space it is still typically the responsibility of Japanese women for performing care. Therefore it is no surprise that older men are rejecting the smartphone; they are behaving in line with gendered norms regarding domestic spaces of sociality that have been established over a lifetime.

Redefining the fourth age

As we have seen in the cases above, interdependency in various forms was experienced among all of my participants at varying ages, and rather than a clearly defined 'third age' followed by a 'fourth age' it is the anticipation of frailty on the horizon that can either motivate people into actions of preparedness or make them shrink back. It is instructive to see how such opposite attitudes can be witnessed even within one couple. This next case demonstrates that, rather than viewing ageing as merely a sequence of ageing categories corresponding to increasing decline, what has been termed the 'third age' incorporates the way people respond to and prepare for issues anticipated in the 'fourth age'. The kind of pro-active attitude most associated with the 'third-age' can emerge precisely when facing issues of the 'fourth age', and this 'final' stage of life can be redefined as a stage of action and preparedness as we will see below with the emergence of the phenomenon of *shukatsu*, or 'end of life hunting'. This finding is important because it challenges the very distinction between the third and fourth age, showing how the future is prefigured in the present.

In my rural fieldsite of Tosa-chō, one of the staff members at the local council was undertaking a project to photograph every person that made up the town's population. At the time of starting the project, this was 4,001 people. This project was met with delight by many of the residents, but for Suzuki Sachiko-san, a woman just about to turn 80, she was hesitant about having her photograph taken by a professional photographer. This hesitancy was not because she was worried about how she would look, but that the act of taking the professional photograph itself would make her more keenly feel her age and remind her of her mortality.

When someone dies, typically their family choose a photograph of them to be used at their funeral and this image is subsequently displayed on the home altar (*butsudan*).[76] Writing about Japanese memorial practices, anthropologist Jason Danely observes that through frequent interaction and visualising techniques at graves and *butsudan,* older adults especially 'recognise the mutual interdependence of the living and the spirits of the dead'.[77] Photos of the deceased play a key role in the re-socialisation of the dead as ancestors which are worshipped as part of Buddhist cosmology. People are increasingly choosing more natural photos, such as from holidays, as memorial images;[78] a photography studio near to my home in Kyoto, however, advertised a package on more formal portraiture that directly targeted an older clientele. The key selling-point was that they offered to edit out wrinkles and to present a customer's 'best side'. Several of my female participants regularly used beautifying filters in their smartphone selfies, adding make-up and smoothing out the skin, and now these digital capabilities were moving to more traditional photographic settings.

The text on a flyer available at the studio's entrance assumed that customers would be fairly used to seeing themselves on screens and would know when the camera had captured their 'best side'.

> We will let you take your time and take as many photos as you like until you are satisfied with good facial expressions showing the true customer. You can see [the photo] on the monitor screen immediately, so you can rest assured.

Digital photographic practices such as photo-editing that have advanced with the smartphone have become part of how important milestones are commemorated throughout the life course and beyond.[79] While this flyer did not explicitly state that the photo could be used for the purposes of a commemorative photo after death, it did so implicitly. The most affordable photo, suitable for a photo album, was advertised as being possible to enlarge for a picture frame at a later date.

For Sachiko-san, having a professional photograph taken in later life was too much of a reminder that it could potentially be used in the event of her death as part of the way she would be remembered by her family. It made her feel sad to think about this movement into the last stage of life and subsequently the realm of ancestors.

> It is scary. Tanaka-san [the council staff member] said he will take photos, and when he develops them I feel that I am going to the other side! My last photo … So I tell him to wait, not to take photos.

Her husband Hiroshi-san had a completely opposite opinion, however. He had specifically asked Tanaka-san to do what Sachiko-san feared most: come to his home and take his commemorative photo. Sachiko-san thought this was brave and they agreed to disagree in their views. Commemorative photography is a key item of material culture in the homes of many of the people I met, particularly in rural Tosa-chō, with portraits of the recently deceased sitting on the *butsudan* where family would also place offerings of food and tea. Photographs of older generations often adorned the walls in the rural households I visited, looking down from near the ceiling on the activities of the home. For Hiroshi-san, taking a commemorative photograph was just one activity that was part of a wider process of preparation that he felt they ought to start soon, given their age. Rather than explain his reasoning in terms of a decline, he said that as they are both nearly 80 years old, they 'are nearing the age of *shukatsu*' – an interesting turn of phrase that plays with words associated with other age-based norms. The term '*shukatsu*' is used often by younger people to mean 'job-hunting', but since the early 2010s it has become a buzzword among older people to mean planning for later life and the afterlife.[80] By swapping the character 'shu' for a similar sounding character, people play with the word to make it mean 'end of life hunting'.

During the early 2010s the term gained popularity in the media; it corresponded with a period in which the first baby boomers reached the age of retirement and there was growing concern about the ageing society.[81] Hiroshi-san included a range of activities in '*shukatsu*', mostly revolving around getting rid of stuff from the home, something that he found far easier to do than Sachiko-san. They also made sure their bank account details and any important information relating to inheritance were easy to access for their children. 'Not that we have any money!' Hiroshi-san remarked with a laugh. While they enjoyed multiple pensions and had enough to live on, they often gave the money they earned through their flower business to their grandchildren. They felt acutely sorry for the younger generations who faced more economic uncertainty than they felt they had experienced in their own lives, even though they had started out with very little. This sense of sympathy for the younger generations made them want to reduce the burden that they placed on their children in their later life.

Both Hiroshi-san and Sachiko-san feel at the cusp of entering a new life stage in which they anticipated, with a mix of resignation and trepidation, a period of potential decline but also a movement towards becoming an ancestor. The refusal of the portrait photograph by Sachiko-san was one way for her to delay entering this stage despite her advancing years, situating her in a liminal period of not young, but not yet old. Whether or not there would ever come a time when she would feel ready to enter this stage remained unclear. Unlike her husband, she prefers to see entering '*shukatsu*' as always just over the horizon, regarding Hiroshi-san as brave for facing the next period of life head-on and being ready for death at 'any time' – an attitude which included having already moved his ancestors' remains to near their house, organising the grave of himself and his wife among them. Hiroshi-san explained that while future generations may forget them, the grave is a location of permeance, linking them both to the past and future.

> Our names will be carved in it. Ancestor after ancestor. We have an old family tree, so it'll be like that. And those things never vanish.[82]

The grave is a more enduring memorial than the commemorative photo. In making such plans Hiroshi-san was ensuring that they were entering their final stage of life with agency and determination, as well as continuation on land that means so much to him.

What such actual life experiences and attitudes reveal are unexpected re-configurations of the academic and official approaches to the categorisation of ageing. Flipping the 'shadowy' fourth age narrative[83] and using it to create a new stage in the life course focused on action and care towards the next generation, Hiroshi-san forged his own definition of successful ageing, turning it into something different than simply the idea of remaining active. Yet these ethnographic vignettes around individuals and their experiences demonstrate in turn the wider context of changes in life expectancy and political economy in Japan.

Like many of my research participants who were from the post-war baby boomer generation (*dankai no sedai*), the Suzukis experienced the rapid economic growth of the 1960s to 1980s, during which a capitalist model of 'life course' was promoted by the state.[84] During this time Hiroshi-san sold his farming equipment, moved to find work in Kōchi city, married Sachiko-san and bought a house. For most men of his generation, the normative life course model included clear age-based stages, such as entering the workforce at the age of 23, marriage and the birth of a first baby by 30, followed by a timeline centred on the development of one's children, ending with one's own retirement at the age of 60.[85] The high-growth era dictated that one's progress through life should be demarcated by a sequence of socio-economically 'productive' moments and the accumulation of middle-class goods.[86] In later life, then, a new life course stage may be emerging in the divestment of these accumulated goods.

The Suzukis are preparing their home to be taken over, or sold, by the next generation in anticipation of their deaths, rather than foreseeing a time in which they may all live together, as with the Katsura family in Osaka. Once again technology plays an important role as a mediator to facilitate closeness without the burden of intergenerational living. Their children live in Kōchi city, an hour and a half away by car, and they use a family LINE group to see photographs of their grandchildren and share the kind of informal sociality that would have come through living together. Sachiko-san enthusiastically scrolls through the photos of her grandchildren on her phone, in direct contrast to her feelings about having her own photo taken by the photographer.

Photographs keep the ancestors close by in the home, a visual reminder of connection and belonging through a lineage of generations often long gone. However, photography today has taken on a new liveliness and significance in keeping living family members close. It is in the smartphone that these images are primarily displayed, not on altars or walls. The physical home as a place where a family dwells is thus reduced in significance, while digital dwelling through visual communication becomes an everyday mode of affectivity and connection.

Conclusion

Using technology to mediate family relationships is not new, nor even limited to those living apart, as we saw in the case of the Katsura household with the intercom. It is rather part of a longer trajectory towards increasing autonomy among multiple generations. One reason behind this desire for connected independence was identified by John Traphagan in his study of how older adults contest their progression to 'elderly' status.[87] Traphagan found that older Japanese adults refuse the term through practices that reduce the trouble they might cause to others (*meiwaku*) by relying on them for care. The cases outlined in this chapter show how such behaviour is migrating online, practised in combination with *shukatsu*, demonstrating the multiple ways which people live with – and against – categories of ageing. This chapter established this book's intention to develop analysis through detailed portraits of individuals or relationships and their often-contradictory experiences and opinions. Such portraits highlight uniqueness while simultaneously demonstrating how people are embedded in wider social contexts and norms.

The vignettes presented above serve to demonstrate how people live in between and against categories of age. Such categories include those that have arisen in popular discourse or through government policy, or indeed through academic engagement, for example 'third' and 'fourth' age[88] and the concept of the life course.[89] Rather than establishing a simple antithesis between categories and experience, this chapter has argued for a more fluid approach to understanding how people respond to categorisation of their life course. Becoming 'elderly' had negative connotations for some people, due to a fear of increased isolation. The smartphone is becoming one way that people resist such categorisation by assisting with social connectedness. My older interlocutors, such as the Suzukis in Tosa-chō, who were entering their eighties, revealed mixed feelings of trepidation and resolve when anticipating their future and were concerned to reduce their burden on others. Meanwhile others, such as Haruki-san, are exploiting the freedom of their advanced age to behave in ways that previously they may not have been able to due to social judgement.

Although some people such as Haruki-san may never accept the category of 'elderly' because of an association with decline, Suzuki Hiroshi-san demonstrates how conceptions of entering the final stages of life can also revolve around activity and agency. Both Haruki-san's refusal and Hiroshi-san's preparedness could be seen as evidence of 'successful ageing' according to the definition originally outlined by Rowe and Kahn,[90] which focuses on the individual's responsibility of rising to the challenge of ageing through 'individual choice and effort'. However, what is clear from the ethnography is that it is not individual action but interdependent relationships of care and responsibility for which people strive. The concept of 'successful ageing' also provokes the binary of 'unsuccessful ageing' through its polarising language and focus on the individual.[91] Yet in the cases presented above, we see how webs of support are interwoven between offline and online, between generations, enabling people to maintain autonomy even when they might need increased care. Such evidence accords with the findings of Iza Kavedžija, whose ethnographic research in Japan suggests that people are situated in webs of interdependency throughout their lives, and that it is through these that they actually achieve independence.[92]

With a rapidly expanded life expectancy emerging within just one lifespan, new opportunities have arisen for many of the people I knew compared with previous generations. The following chapters will detail how milestones of later life, including, for example, gaining post-retirement employment, volunteering, enrolling in education or losing a spouse, marked transitions in similar ways to earlier life course moments.[93] However, Keiko and Takashi Katsura in Osaka feel that the generation above them is special in terms of their longevity. Like many people I spoke to in their sixties, they do not believe that they will live as long as their parents' generation, born in the early *Showa* era of 1926–34 and with childhoods and formative years that spanned the hardships of the Second World War.

In the popular imagination, the baby boom generation is comparatively lacking in moral character, being primarily focused on personal lifestyle.[94] The Katsuras feel that the lifestyle and diet of older people is healthier than their own and that of younger generations. They feel that the super longevity that currently defines discourses of ageing in Japan is not necessarily a given fact for the future. This scepticism only emphasises that conceptions of ageing are open to change and dependent on the context of the day. It is wrong to assume that the ideals to which many people in this research aspire, to stay connected and to reduce burden, will be the same for the younger generations as they age.

While most people who appear in the rest of this book were in their fifties, sixties and seventies and could be considered 'pre-old' according to official categorisation, the difference between them and those in their eighties and nineties presented here was not clear cut. Throughout, differences between individuals were more prominent than differences between age cohorts. Some people worked hard to avoid physical markers of age; others were more ready to accept and identify with their changing bodies. The diversity represented just among the people I met was immense. Categories such as 'third-age' or 'pre-old' gloss over the lived reality of older people, who are continually experiencing forms of dependency, decline and, conversely, blossoming and independence.

Even when people do experience increased need for physical care in the so called 'fourth age', they are not suddenly without agency; indeed many are also performing care towards others, even if the expression of that care is an appreciative comment on a photo shared digitally. As anthropologist Nikita Simpson notes in relation to her work with older rural-dwelling women in Himalayan India, 'care for the elderly is not only physical care or even respect, but a wider ethics of acknowledgment, attentiveness and appreciation'.[95] It is this broader sense of care as appreciation that I found to be emerging most strongly in the digital sphere via communication across and within generations.

In the ethnographic examples above, the home emerges as a key site for the negotiation of agency, autonomy and dependency. Through emerging digital practices of care at a distance,[96] we see how the smartphone has become a space into which these concerns extend, and can thus be regarded as an extension of the domestic space itself. The ethnographic work undertaken by anthropologist Inge Daniels within the Japanese home demonstrates how people carefully manage material and spatial practices to create a domestic atmosphere that is warm but not overbearing, striking a careful balance of individual autonomy and collective dependency.[97]

I have introduced here a key theme of this book, which will be developed in subsequent chapters. It considers how visual digital communication is emerging as a key material practice through which people foster intimacy as well as distance with the people in their networks of care. Sharing photographs of meals, for example, allows a mother and son to engage in moments of domestic connection centred on food even while living separately.

The smartphone thus emerges as a key technology for how people stay embedded as social persons in later life. However, we have also seen how this labour can be heavily gendered, perpetuating the responsibility of women for care and emotional labour. The relationship between visual digital communication and 'emotion work'[98] will be explored in Chapter 5, along with a discussion of how gendered practices of care are emerging through the smartphone. The emergence of digital forms of care provokes questions about who may be marginalised as a result. A lack of digital skills or the inherent ageism and ableism in the design of many digital devices and services may in fact lead to a withdrawal of care. As many people stressed to me, the smartphone cannot replace face-to-face care. However, they were equally enthusiastic about how it can facilitate caring relationships, reducing feelings of burden and opening new possibilities for agency and sociality, two of the things that I found people most fear losing in later life.

Notes

1. A pseudonym.
2. Kavedžija 2015b, 77.
3. For example, see Chapman 2005, Portacolone 2011, Rubinstein and de Medeiros 2015 and Lamb 2014.
4. 'Successful ageing' as defined by Rowe and Kahn 1997. For example, see Danely 2019, Kavedžija 2019, Lamb 2017 and Fujiwara 2012.
5. Moore 2017.
6. Fujiwara 2012.
7. Rowland 2012, 222.
8. Rowland 2012, 222.
9. Following Kavedžija's (2015a) assertion that older adults in Japan are engaged in a series of 'balancing acts' such as between sociality and over-closeness, dependence and autonomy, in their pursuit of happiness.
10. Ikeda et al. 2011.
11. Human Mortality Database 2021.
12. Tsugane 2021.
13. Ministry of Health, Labor and Welfare 2020b.

14. Ogawa 2015.
15. Suzuki 2012.
16. Moore 2014.
17. Pickard 2019.
18. This motion was proposed by the Joint Committee of the Japan Gerontological Society and the Japan Geriatrics Society, as described by Ouchi et al. 2017.
19. These figures correspond to a census conducted in 2014, which revealed that many citizens identify old age as starting at 70+ (30 per cent) for men and 75+ (29 per cent) for women. The next highest percentage of respondents answered that old age is in fact not decided by chronological age at all (10 per cent) (Cabinet Office of the Japanese Government 2014).
20. Terms that are commonly used to refer to older adults in Japan include *kōrei-sha*, which means 'elderly person' and is used in official discourse. The term has connotations of incapacity, for example the 'Kōreisha mark' is a statutory sign in the Road Traffic Law of Japan to indicate 'aged person at the wheel'. Anyone who is aged 70 and over must place the symbol on the front and back of their car if it is deemed that their age could affect their driving, and drivers aged 75 and over have to display the mark by law. The mark, in the shape of a four-leaf clover, was originally in orange and yellow hues, denoting an autumn leaf. Now, however, the symbol also includes two tones of green in a nod towards elderly vitality (Davies 2019).
21. *Shinia* (meaning 'senior') is an English loan word; it serves as a more neutral term that can refer to anyone over the age of 50, but typically to people in their sixties and above. Another English loan word which has become ubiquitous is *siruba* ('silver'), often added in front of other words to denote services for silver-haired people. Other words for older adults include *rojin* ('elderly people') or its alternative *jitsunen*, proposed by the Japanese government in the 1980s as a more positive term; it translates as 'age of fruition', but has failed to become widely established (Backhaus 2008). The term *chunen* directly translates as 'middle-age', which typically refers to people aged between 40 and 60, though even 40 years ago research suggested that people at the lower bounds of this age range preferred the term *sōnen*, meaning 'prime of life'. The term *chōju* (longevity) has also come to prominence since the 1980s with the rise of Japan's ageing society. It is often used in connection with the promotion of health among older people. For example, in the Kōchi Prefectural office I became acquainted with a department called Office for the Promotion of Longevity. This department had recently designed an app aimed at improving the health of older adults in the prefecture, which I will discuss in Chapter 6.
22. Ouchi et al. 2017.
23. Laslett 1991; Gilleard and Higgs 2002.
24. Higgs and Gilleard 2015.
25. Gilleard and Higgs 2010; Higgs and Gilleard 2015.
26. Gilleard and Higgs 2011; Pickard 2019.
27. Gilleard and Higgs 2011; Pickard 2019.
28. Van der Horst and Vickerstaff 2022.
29. Verbruggen et al. 2020.
30. Bartlett and Kafer 2020, 255, as cited by van der Horst and Vickerstaff 2022.
31. Giele and Elder 1998, 22.
32. Danely and Lynch 2013.
33. Danely and Lynch 2013, 4.
34. Plath 1980.
35. Izuhara 2015.
36. Danely 2015a.
37. Hutchinson 2018.
38. Cook 2013.
39. Pina-Cabral and Theodossopoulos 2022.
40. As described in Mannheim's classic essay 'The Problem of Generations', Mannheim 2023 [1928].
41. Mannheim 2023 [1928]; Edmunds and Turner 2002.
42. Kelly 1993a.
43. Kelly 1993a.
44. Sakaiya 1976.
45. Ishikawa 2020.
46. Miyamoto 2014.

47. Ishikawa 2020.
48. Ishikawa 2020.
49. Kavedžija 2019.
50. Mueller and McCollum 2022.
51. Takahashi and Danely 2020.
52. Takahashi and Danely 2020.
53. Pina-Cabral and Theodossopoulos 2022.
54. Cabinet Public Relations Office 2017.
55. Allison 2013.
56. Many older people in my research tended to retain their *garakei* (feature phones) for cost reasons. This behaviour will be discussed in Chapter 5 on smartphones.
57. Kyoto City official website 2021.
58. The *ie* system, whereby a woman upon marriage enters into the household of her husband (*yometori* marriage), became typical in the eleventh or twelfth century (Wakita and Phillips 1993). Mackie argues that the most conservative form of *ie* – a patriarchal family based on primogeniture – was codified in the Meiji Civil Code of 1898, despite a variety of marriage and inheritance practices that had existed prior to the Meiji restoration of 1868 (Mackie, 1997). The term still has significance today, even with changing living arrangements, since the family registration system (*koseki*) based on a household with a head of the family, revolves around this traditional principle. Along with inheriting property, the eldest son (or, more accurately, his wife) also inherits the responsibility to care for ageing parents. The notion that care is part of domestic work and as such is the responsibility of women remains, despite the increased participation of women in the labour force outside the home (Mackie 2015).
59. 奥さん (*okusan*) is the most common term for wife in Japanese. The kanji 奥 (*oku*) indicates the inner part of the house, historically the wife's domain.
60. Long et al. 2009.
61. Tsutsui et al. 2014.
62. Traphagan 2006.
63. Traphagan 1998; Danely 2013; Kavedžija 2019.
64. Ogawa and Retherford 1993, 91.
65. Ogawa and Retherford 1997.
66. Knight and Traphagan 2003.
67. Hirayama and Izuhara 2018.
68. Daniels 2010.
69. I will pick up this thread later in this book in Chapter 5 on smartphones. There I describe a community-wide broadcast system located in many homes in rural Kōchi.
70. Daniels 2015.
71. Daniels 2015.
72. Nemoto 2013.
73. Hendry 2017.
74. White 2018.
75. Borovoy 2005.
76. Danely 2015a.
77. Danely 2015a, 23.
78. Daniels 2022.
79. These include 60th birthday (*kanreki*), *kōki* (70 years old), *kiju* (77 years old), *sanju* (80 years old), *yoneju* (88 years old), *sotsuju* (90 years old), *hakuju* (99 years old), *momoju / kiju* (100 years old). Other photo studios went even further, advertising their services for marking even later ages such as *chaju* (108 years old) and *koju* for when one reaches the age of 111.
80. Chan and Thang 2022.
81. Swane 2017.
82. Much like the commemorative photo, a piece of material culture connecting family members through time, family graves were important in Tosa-chō. One man had eight generations of graves on his plot of land. I will discuss the importance of a connection to land and rootedness as people age in Chapter 7 on rural migration.
83. Higgs and Gilleard 2015.
84. Kelly 1993a, 201.
85. Shakuto 2017.
86. Kelly 1993a.

87. Traphagan 1998.
88. Gilleard and Higgs 2010.
89. Giele and Elder 1998.
90. Rowe and Kahn 1997, 37.
91. Rubinstein and de Medeiros 2015.
92. Kavedžija 2018.
93. Elder 1998.
94. Kelly 1993a.
95. Simpson 2022, 406.
96. Pols 2012.
97. Daniels 2015.
98. Hochschild 1979.

Image by Laura Haapio-Kirk.

3
Everyday life: gendered labour and non-retirement

> Everything is done now. I've accomplished everything. By writing now, it feels good to live. Writing gives me power, power for life. When we age, it's very sad. Old age feels lonely … sadness. Because Japanese women work so hard to raise their children, their hobbies are out of the question. They fully commit to their husbands. I have lived this way for 40 years, concentrating on my family. I am alone now and my children are independent. It's finally my time.
>
> Hideko-san, aged 70, Tokyo.

Hideko-san is a 70-year-old woman who had recently moved to Tokyo from Osaka to pursue her dream of becoming a professional screenwriter. She had previously dedicated herself to her family, but always knew that she wanted to find a way to express herself through writing, that she had stories to tell, but she could never find the time to develop this passion. After her husband passed away and when she reached the age of 70, Hideko-san decided this was the right time to take action and enrol in a screenwriting course.

At the same time Hideko-san and her daughter were managing the property business that her husband had built up. This business allowed her the freedom to pursue her new career in her 'retirement' from domestic work. Many women I met of Hideko-san's age were similarly driven to pursue their goals after a lifetime of looking after other people by starting businesses and finding work outside the home. These urgent desires to seek new kinds of experiences and careers were motivated by a realisation of how gendered labour norms had previously limited their opportunities for self-fulfilment outside of raising families.

> I love concentrating on my writing. I write my stories and then read them and get so emotional! Crying! I feel so much sympathy for the characters and feel angry for certain characters! I sometimes am

on the metro and if I see a boy that I like then it sparks inspiration for me and I want to write someone like him in my stories! ...That makes me happy because I feel the same age as my characters.

Through her writing, Hideko-san felt less lonely. She also felt that she transcended her age. Discovering YouTube on her smartphone gave Hideko-san the motivation to publish her own short films on the platform one day; to do this, she intended to hire actors to perform her screenplays. It was all planned out. A central motivating force behind Hideko-san's ambition to write was to make social connection with others – whether they be fellow students on her writing course, the characters in her stories or the audiences she might reach in the future. Her daughter Yuriko-san supported her mother's dream by leaving her job and making the move from Osaka to Tokyo. Yuriko-san explained that this huge upheaval in both of their lives was worth it because her mother had discovered her purpose in life (*ikigai*), whereas she was still searching for her own. Yuriko-san was thus happy to defer to her mother's dreams for now. Yuriko-san, aged 43, is unmarried and has no children, yet her action mirrors the kind of devotion and self-sacrifice that a woman of her age traditionally made for her husband and children.

> I think when she's writing she can forget her age. She is divorced from her age ... I got a message from my mother saying she might want to skip school today, and I was in Osaka but I said 'No!! You must go!!!'

This kind of role reversal may become increasingly common in Japan as more women delay having families, while the older generation finally have the time and money to fulfil ambitions they always dreamed of, requiring the support of those around them. Chapter 8 will delve more deeply into the topic of *ikigai* and its gendered and generational inflections. First, however, it is important to grasp how the centrality of continued work in later life is construed differently for men and women. For Hideko-san, pursuing her dream career was only possible in her seventies, once duties of care towards her family had been fulfilled.

It is impossible to talk about labour, or indeed anything, in Japan without acknowledging the stark gender divides that continue to dominate contemporary life. Housework, childcare and care of those nearing the end of their lives have typically been the domains of women; many of my female participants had withdrawn from the workforce when they married or had children. The gendered division between household labour and paid work is not unique to Japan, of course, but the country continues to have particularly high rates of withdrawal from paid work by women upon marriage and childbearing.[1] This situation has deep implications for how older men and women then approach labour in later life.

This chapter discusses the lived experiences of the inequalities of irregular work which disproportionally affect women. However, the women I met also found social connection and opportunity in such work, despite its systemic problems. Most recently, Japan ranks as 121 out of 153 countries in the World Economic Forum's 2020 Global Gender Gap Index. Former Prime Minister Abe's 'womenomics' policy is thus clearly not achieving its goals of gender equality.[2] Yet, rather than thinking of gender asymmetries as determining the everyday lives of women along narrow, pre-determined paths, the ethnographic vignettes presented in this chapter speak to an intriguing diversification of experiences of everyday life. Today both men and women must find their own ways of navigating wider pressures and inequalities.

In this chapter I show how people are motivated to continue forms of work beyond the official age of retirement by a mixture of self-interest and social expectation. My research participants often experienced a combination of a desire to fulfil personal interests with the wish to prove themselves as self-reliant in a societal context in which this was praised. The recessionary era and associated neoliberal policies in Japan[3] have meant that a continuation of work can be seen as a way to maintain health and independence for one's self at the same time as acting altruistically for the benefit of others. Such an attitude is part of a moralising discourse around ageing that places responsibility on older adults to demonstrate care for others. Yet this drive towards staying active through continued work is also motivated by personal desires, for example to relieve boredom and an uneasy sense of being suddenly unproductive after a life often dominated by work, the latter experienced especially by men. In order to understand the place of work in older people's lives, it is first necessary to present a brief historical context of paid and unpaid labour for both men and women in Japan, to which the chapter now turns.

Work and retirement

Maintaining participation in paid work in later life is not limited to Japan. It is common among other societies with ageing populations, such as the United States,[4] and has led to the raising of retirement ages in many developed countries.[5] Greater labour force participation among older people is a priority of the government in Japan,[6] a country with the highest old-age dependency ratio (OADR) in the world.[7] The total labour force of the country has been in decline since 1998, yet the rate of participation in the workforce among people aged 65 and over is high.[8] In 2022, for those aged 65 to 69, the average percentage of people in employment was 61 per cent for men and 41 per cent for women,[9] compared to just 10 per cent in Europe.[10] The increasing participation of older people, particularly men, in the labour force over the past 10 years is in large part due to the effective raising of the pension age to 65 as well as a revision to the law in 2012; employers are now required to employ workers until the age of 65.[11] However, many of my research participants between the ages of 60 and 65 were working reduced hours, often at their employers' request.

In April 2021 the Cabinet effectively raised the retirement age again from 65 to 70 through bills requiring companies to retain workers,[12] but the pension age of 65 remained the same in principle.[13] The national pension plan includes the option for individuals to defer receiving a pension from 65 until any age they want, when they are rewarded with a higher rate of pay. With working life extended, participation in social activities outside of work reportedly decreases if one does not begin such activities by the early seventies – highlighting the need to understand the implication of later retirement for the way in which older adults participate in communities outside of work.[14]

The demographic shifts of a shrinking and ageing population have had a significant effect on the Japanese labour force. When walking around Kyoto, it is striking how many workers employed in jobs such as directing traffic around building sites or building maintenance staff are in their sixties and over. The low birth rate and corresponding labour shortage, combined with low levels of immigration, mean that there are many essential jobs which the newly defined 'pre-old' population are in a prime position to fill. As discussed in Chapter 2, the proposal to change the bureaucratic definition of those aged between 65 and 75 to 'pre-old' is in line with a proposal for legislation that would allow companies to employ workers until the age of 70.[15] The recent rise in participation rates in the labour force among older adults is a trend observed not only in Japan, but also in other developed countries, many of which are also undergoing processes of pension reduction.[16] However, the continuation of work into later life has different implications for men and women in Japan. This is not surprising given the gendered labour norms that determine the experience of everyday life for many.

Gendered experiences of labour and the crisis of care

Historically the gendered labour dichotomy was rooted in state ideology that sought to establish modern, industrial Japan, and indeed Japanese-ness itself, as masculine.[17] The modernisation of Japan from the late nineteenth century saw an intense focus on the home and community as sites for establishing social values conducive to industrialisation. Official discourses of women as 'good wives and wise mothers' (*ryōsai kenbo*) emphasised the female role to be almost a kind of civil servant within the domestic sphere, doing their national duty by raising healthy children and supporting their wage-earning husbands.[18] The post-war decades saw the labour of women emerging primarily in irregular forms of employment. This gender split continues to shape the division of labour in homes and to mediate the possibilities of what both men and women can be in the post-bubble era.[19]

During the high growth era of the 1970s and 1980s, when many of my participants were starting their families, domestic and care labour was still highly associated with femininity.[20] From the 1980s the participation of married women in paid work rose, accelerated by the passing of the Equal Employment Opportunity Act in 1985.[21] However, the increase has largely only been in part-time or irregular positions, as women balance paid employment in addition to housework and caring.[22]

In the twenty-first century many younger women do not intend to quit their jobs upon marriage, while older women are returning to paid work rather than volunteering in neighbourhood associations, typically important institutions for social care.[23] However, being in full-time work does not diminish women's domestic duties, which they now have to balance with paid labour. Problems associated with the rapidly ageing society and declining birth rate have been closely tied to the Japanese government's policies regarding gender equality, which have received criticism regarding their methods and goals, such as from feminist theorist Kano Ayako.[24] Kano argues that Japanese gender policy is grounded in, and thus reinforces, the idea of all women being 'potential mothers',[25] narrowing the possibilities for what it means to belong to the category of 'woman'. However, with increasing numbers of women not having children in Japan, policy seems out of step with the lived realities of women, many of whom will not be able to rely on the kind of care that they currently provide for the generation above them. As this next ethnographic example will demonstrate, responsibilities of care sit alongside commitments to work; both are implicated in moralising around duty and lineage.

Misako-san in Kyoto is in her early sixties and has never married. From a young age she felt sure that she would never have a typical family life, remembering as a young girl how her grandmother had been told by a fortune teller friend that Misako-san would not marry, and thus it came to pass. As the eldest of two sisters and the heir to a large family business, Misako-san knew that it would be her responsibility to continue the efforts of her father with the company; in a way she always felt more like a son than a daughter. A male heir had been the wish especially of her grandfather, and she recalled how as a young girl he would call her *bōya* (boy).

Misako-san now works long hours, six days per week, and often takes her work home with her. Feeling increasingly isolated in her work, especially as a female boss in a male-dominated environment, she admitted to me one day that 'I hate business. Money brings out the dark side in people'. Often seeming troubled by the pressure she is under, she confided that she sometimes feels trapped and exhausted by the situation. Because she had no choice in running the business, she states that work is definitely not her purpose in life (*ikigai*); implicating choice as an important factor in what constitutes *ikigai* – something that I will consider further in Chapter 8. However, since she is so concentrated on her work, she cannot think about her life outside of it.

Despite being in her early sixties, Misako-san sees no possibility of changing her life any time soon. She is waiting for her mid-sixties to 'retire', i.e. to reduce the number of days she works. One day when we were in an art gallery, she particularly liked a drawing of a penguin floating on the top of a wave, its webbed feet paddling away below the surface. 'It's just like me,' she said, meaning that although on the surface she may appear calm, underneath she is frantically paddling to ride the waves.

The responsibility of the family business is not the only pressure on Misako-san's shoulders. As discussed in Chapter 2, the Japanese model of a Confucian welfare state relies on the family and community as primary welfare providers. In recent years the government have introduced much debated measures to take some pressure off women as caregivers. Despite these efforts, the Japanese welfare state continues to rely significantly on the unpaid labour of women within families. Misako-san's mother is now in her late eighties and lives alone in their large family home, which was rebuilt in a modern style. They sometimes attend events together at the temple and Misako-san will often bring groceries to her mother's house on a Sunday. She lives in an apartment a short drive away and is available to care for her mother when needed, while maintaining a desired level of distance. Misako-san and her sister are involved in arranging her mother's hospital appointments. When we visited her mother together, Misako-san confided that she had been struggling to get her mother to agree to have the hip operation she needed, asking me whether I could help to persuade her.

Misako-san observed how in recent years there has been an increasing reliance among many Japanese people on social services. This dependence, she believes, stems from various factors, including a decline in the number of children people are having and a rise in single-person households, much like her own. The shifting demographics include a growing number of single women and men who often find themselves solely responsible for the care of their ageing parents. With the necessity to maintain employment, single people frequently turn to social services to provide care for elderly relatives, a task that would be otherwise unfeasible. It is evident to Misako-san that Japan is undergoing a gradual transformation. Despite the persisting inequality between men and women in society, changes are perceptible, albeit at a slower pace, she feels, compared to other industrialised societies. Social service vehicles offering day care services for the elderly are frequently seen around Kyoto, providing a visible indicator of the evolving dynamics of care.

When we visited Kitamo temple together one day, Misako-san, her mother and I sat tightly packed on the floor with hundreds of others to hear a guest sermon by a visiting monk. The key takeaway from his hour-long talk was that you should not say to other people 'I'm so busy' (*isogashi*) because it will make them feel guilty for taking up your time. Misako-san commented that it depended on how a person might say it, and to whom. She felt it was quite common for people to imply how busy they were indirectly. The hall was filled with people in their sixties and over who, as my research clearly revealed, were indeed likely to be very busy. In particular women such as Misako-san, who were nearing retirement age or engaging in post-retirement work, probably also faced increasing care demands of ageing relatives. The moral imperative for women to take on the labour of care, and to not express or even acknowledge this burden, as suggested by the monk, is deep-rooted in social policy. As Kano states:

> The basis of social policy in modern Japan is the assumption that all women are potential wives and mothers (and that all men are potential breadwinners and heads of households). Almost everything else derives from this assumption.[26]

Misako-san is glad that her mother now enjoys a comfortable life, conscious that she faced a difficult time when she first entered her husband's family home as a newlywed in the south of Kyoto. She has vivid memories of her mother having no money of her own, and of her having to ask her father's mother for house money if she wanted to do anything with her children. One day, when the film *The Sound of Music* was newly released, Misako-san's mother found the money to take her and her sister to see it. They enjoyed it so much that instead of watching the film once, they stayed in the cinema all day to watch it multiple times to get their money's worth. It was dark when they finally emerged from the cinema to go home.

In comparison, Misako-san as a single person is bound by different constraints to her mother. Her gender remains a significant factor in her everyday life, both in how she juggles the responsibilities of business and care and in how the competing demands are intertwined. I would argue that the 'crisis of care'[27] that affects women in many parts of the world is heavily shaped in Japan by the moral duty of filial piety. Such a duty not only influences commitments to caring for one's family, but also encompasses extended notions of ancestral lineage and heritage.

What struck me as Misako-san and her mother leafed through old albums and talked about their family history was how this once large family was now reduced to Misako-san, without children, and her younger sister, whose son wanted no involvement with the family business. The pressure of continuing the family story rested on Misako-san's shoulders alone. Pointing at the family altar (*butsudan*) in the corner of the room, she told me how it had been rescued once from a house fire – the only item to survive, it had been passed down through the generations. This large and heavy object – with its beautiful and ornate gold display – was a physical manifestation of the very real and heavy responsibility passed down to Misako-san for upholding her father's business since he passed away.

For Misako-san, as with other older business owners and craftspeople I met in Kyoto and elsewhere, passing down family trades is now often impossible. Either a younger generation does not exist, or it has different aspirations and attitudes about work and responsibility. Misako-san's commitment to her filial duties involves an intense feeling of care and connection towards the past, perhaps precisely because of the lack of a younger generation to take the business forward into the future.

A steadfast maintenance of traditional crafts by individual practitioners in the face of a lack of generational continuity was evidenced by another Kyoto craftsman I met. He was from a line of 11 generations of water pot makers, used in the traditional tea ceremony. The market for these expensive items is shrinking due to Japan's stagnant economy, and he anticipates that his long-standing family craft will disappear after his death; it is not financially viable for his children and grandchildren to continue. Declining birth rates and the precarious economy significantly impact the continuation of traditional family trades, making it even more difficult for older adults who see themselves as the 'last generation' working in a proud tradition to retire.

'Silver, do your best!'

Another form of moral imperative driving the continued commitment to work in later life concerns the very nature of identity and sociality among older adults in Japan. As anthropologist Katrina Moore argues, Japanese notions of personhood have historically been understood as embedded in social relationships, yet are increasingly framed by the concept of personal responsibility (*sekinin*), self-reliance (*jiritsu*) and accountability as a person belonging to society.[28] For those experiencing retirement during this contemporary time of shifting values and ideas, the very notion of what constitutes a person is in question. Anthropologist Nana Gagné Okura's ethnography among company workers in Tokyo argues that the Japanese 'social person' is constructed through labour.[29] The concept of *shakaijin* (literally translated as 'society person') can be understood as fundamental to becoming a responsible adult.[30]

Such self-conceptions are not limited to people entering working life. Indeed, the notion of *shakaijin* as definitive of personhood continues to be important throughout the life course and beyond the age of retirement. Gagné states that 'being a *shakaijin* is about participating in some activity with wider social significance'.[31] Recent research on working identities has focused on the relationship between *shakaijin* and masculinity,[32] but women as well as men are implicated in this identity. The next ethnographic example demonstrates how maintaining a *shakaijin* identity is also bound up with conceptions of health and care, which themselves have social implications.

Keiko-san, the 64-year-old woman whose Osaka home featured in Chapter 2, worked part-time as an administrator in a hospital. Her life did not change much when she reached the age of 60, a milestone year known as *kanreki*. At this time of life, having completed 12 cycles of the Chinese Zodiac, one's age is considered to have returned back to zero. Traditionally at 60 a person might thus have returned to a state of dependence, typically upon their children.[33] Keiko-san observed that 'at sixty I became a baby again', referencing the association between this auspicious age and a return to dependence that mirrors the dependence of infancy.[34]

However, apart from reducing the hours that she worked to include several half-days per week, this milestone was out of step – not only with the official age of retirement at 65, but also with her conception of self-responsibility. Keiko-san wanted herself and her husband Takashi-san to remain working, mainly out of a belief that this was the healthiest way to spend one's later life. 'If he stops work now, he will just watch television all day. It's not healthy!' she lamented. In line with the ethnographic findings of anthropologist Shiori Shakuto's work among Japanese retirees in Malaysia, the motivation behind continued employment for many people I met was to delay the onset of senility (*boke bōshi*), often out of a deep sense of responsibility to younger generations and the desire to not become a burden.[35] Shakuto argues that such a concern is directly linked with neoliberal ideologies that promote self-responsibility and mutual support in retirement. My research points towards a range of reasons for wanting to continue work in later life, many of which focused on self-identity.

When I first met Keiko-san she was still a year away from retirement, but had already intended to find a part-time job through the local Silver Human Resource Centre (SHRC) and had plans to use her days off to take courses at an adult education college. Takashi-san, who was of a similar age, shared her vision. They expected that they would have little choice in the kind of work they would be able to get, and were prepared to take anything that the employment office could offer. Takashi-san hoped he could continue to work for the same hospital that currently employed him – perhaps as one of the people who visit companies and do annual checks on their employees, taking weight and height measurements, as this type of job would get him away from his desk. Alternatively, he conceded with a chuckle, he might end up working as a cleaner in the same hospital.

Silver Human Resource Centres are nationwide in Japan. They were established in the 1970s to place retired people in temporary, part-time jobs that are outside of the traditional labour market, therefore avoiding placing retirees in competition with younger workers.[36] Jobs listed by SHRCs typically pay the minimum wage and include such roles as playground supervisors, home care helpers and clerical workers. The SHRC advertised in the brochure below (Fig. 3.1) is in Osaka City and caters for around 10,000 members. The brochure provides illustrations of the kinds of job that the centre can provide, such as cleaning, care work with older adults, gardening, delivery of flyers, etc.

Figure 3.1   A photo of the Silver Human Resource Centre brochure. In English the title reads 'Leave that job to me!' and the subtitle 'Silver, do your best!' Image shared with permission of anonymous research participant and the SHRC.

When they reached the age of 65 Keiko-san and Takashi-san registered at the centre and took a computer test, examining their skills in Microsoft Word and Excel. After completing the exam successfully, both were hired to work for the actual centre rather than being allocated external work, due to their digital skills. While the SHRC appeared to value their familiarity with computers, Japanese businesses and governmental organisations are often low-tech, favouring paperwork and face-to-face interaction rather than digitalisation – an issue that has been associated with Japan's bottom ranking in labour productivity among G7 countries.[37] This resistance to the digital in the workplace is especially evident in post-retirement employment which tends towards lower-skilled work, as shown in the brochure from the SHRC. The Covid-19 pandemic has pushed the acceptance of digitalisation and telework forwards, but only marginally compared with other countries.[38]

The Katsuras were among the youngest members of the centre. Most were aged between 70 and 75, though there was no upper age limit. 'The purpose (of the centre) is to contribute to the creation of a vibrant community as well as to enrich people's lives,' declared Keiko-san after joining, emphasising how employment through the centre was for social as well as personal benefit. Despite such aspirations, she added that most jobs offered by the centre are cleaning work. Keiko-san's job is to register new members and her husband's is to seek new places, such as domestic or commercial buildings in need of cleaners, in which many of their members find employment. Often members may have been senior company employees in their previous working lives, so they were now experiencing a radical shift in the kinds of labour they were engaged in. Remarking that some of the members were in their eighties and made her feel young, Keiko-san said 'I respect them and want to be as energetic (*genki*) as them'. These older members were role models for Keiko-san because of the way that they embodied health in their adaptation to new working roles and challenges.

Keiko-san and her husband were motivated by the desire to have 'fulfilling days' and to be active members of society 'even at our age'; they thus combined personal and social determination in their reasons for joining their local SHRC. The language used by the centre, with its exhortations to 'do your best' and 'contribute to society' (Fig. 3.1), therefore appears to be aligned with its members' desires. This discourse can be argued to be exploited in turn by a wider neoliberal morality of care that places increased responsibility on older adults in the face of the swelling costs of an ageing population to the state. SHRCs can be understood as a key mechanism through which the state capitalises on the moral imperative to work to offset some of the economic burdens of ageing.

The procurement of care work for older adults by older adults through the Centre was a concrete example of this trend. As anthropologist Shakuto Shiori argues, 'care responsibility in the twenty-first century has shifted from the market to senior citizens themselves'.[39] More broadly, however, it was participation in work itself that was seen as an act of care by Keiko-san and Takashi-san towards themselves and therefore also towards their children – and, more broadly, to society. Motivations behind continued employment in later life tended to be a complex moral imperative that combined individual choice with social responsibility, shaped by wider entrenched employment systems. A deep-rooted sense of selfhood as determined by work became apparent in the anxiety that can arise when both men and women take steps towards a different kind of life in retirement, as shown in the next section.[40]

Fear of change

Stepping away from the world of work was difficult for many people, even if they had tentative, or even well-developed, ideas of what they might do in retirement. People were often keen to find other ways of maintaining an identity as a fully-fledged member of society, or *shakaijin*, possibly through volunteering or other forms of labour that were aligned with their interests. Kawase-san, for example, a 66-year-old male nurse from Kyoto, had tried retirement briefly the previous year, but found that it was not for him.

> I retired at 65, but I am still working. I stayed at home for three months when I first retired and I thought that if I have enough time I could do something very big, but I couldn't. I want to write a novel and I had time, but I didn't. I did some research only.

For Kawase-san, stopping work and the increase of free time this entailed led to a kind of facing up to his life, unconstrained by the demanding schedule of work at a busy hospital. This period of reflection made him realise that he preferred being busy.

> Stopping allows you to look at your daily behaviour. We are living our daily life automatically. When I retired, I felt uneasy. Too much free time made me anxious. Because others are writing novels or being productive, I started to see them as enemies. I could not stay at home. I felt like I have to be somewhere, go out and see people. I had to have something to do. Now I am back at work I have good conversations with co-workers and get new ideas. My mind is stimulated at work.

The imagined productivity of 'enemies' was a kind of unbearable competition for Kawase-san. In contrast, the daily stimulation provided by his mainly younger colleagues provided the kind of sociality that helped him to feel that he was keeping his brain active, rather than representing the threat of competition. Daily interaction with people is what made Kawase-san feel productive; he could not relax with the prospect of all his free time stretching before him. The presence of time constraints had always been important to him for his sense of productivity. Waking up at 2 a.m. was normal for Kawase-san; he had done this for the past 30 years, valuing the morning hours before work as his own creative time. As a single man without living parents and having little contact with his siblings, he was able to dedicate his free time to personal projects. The early morning hours were his time for reading, making games and doing anything else that took his fancy. When in retirement he experienced a complete lack of time constraints, he felt that he was not making best use of the time available to him; there was no work schedule to help structure his days. Now that Kawase-san was back at work he wished that he could only work for five days a week, but his boss demanded that he work for six days. However, despite his complaints about this situation, this struggle to make time for his life outside of work was precisely the kind of tension he desired and needed in order to feel satisfied with his life. Time constraints and interaction with other people were a kind of productive friction, which fuelled Kawase-san's creativity and sense of purpose.

This affective experience of unease and anxiety provoked by a withdrawal from society, or a transition away from being a *shakaijin,* was very personal for Kawase-san yet could be understood to affect many Japanese men whose identities are so bound to their working lives. However, his story parallels the experiences of Sachiko-san, a woman also in her mid-sixties and living in Kyoto. Sachiko-san typically worked six days per week as a teacher and took no holidays. She was careful with money and was known as 'coupon lady' by her friends, since she always had a discount coupon carefully cut out of a newspaper every time we went out for a meal. We would often meet on Sundays and go to galleries or scenic spots together on her only day off. She would frequently tell me how she had been up until 2 a.m. working from home the night before. 'I am a typical Japanese person: work, work, work!' she would quip.

The need to work hard had been established early in her life. Upon learning of her pregnancy outside of marriage, Sachiko-san's father had made it clear that she would find no support from her family. As a single parent with no family support, she embraced the self-identity of someone who worked to excess. This identity, that she felt closely aligned with a national stereotype, would prove hard to let go of for Sachiko-san even after her 'retirement', which finally occurred after I had left Japan.

The rural home of Sachiko-san's parents in Fukushima Prefecture had been empty for several years since her mother and father had passed away. Upon retirement, she decided to leave Kyoto, where she had spent the majority of her working life, and move into this house. Its location was convenient since it was nearer to Tokyo, where her adult daughter now lives. Sachiko-san was looking forward to being closer to her daughter and also to Tokyo, partly in order to fulfil her dream of volunteering for the Olympic Games, which she was indeed able to do when they took place in 2021. Her letters to me during the pandemic mourned the fact that the Olympic Games had been postponed; when she finally got to fulfil her ambition, she sent me photos via LINE of her outside the Japan National Stadium with a big thumbs up. Sachiko-san imagined a 'healthier life' living in the countryside, somewhere she could work less and have time to do the things she wanted like volunteer. However, she also anticipated with some trepidation the lifestyle adjustment involved in her move to the countryside, joking that maybe there she might 'drink a lot and sleep all day – disgusting, right?', as she would say with a laugh.

When Sachiko-san made the move and 'retired', she in fact retained her connection with her employer. It transpired that the nationwide company needed a teacher to run classes at a school located a 3-hour train ride north of her new location. They were unable to find a local teacher, so Sachiko-san agreed to make this round trip several times per week to continue her employment on a part-time basis. Despite the long commute, she enjoyed interacting with the students and was glad to still be working, in combination with her volunteering. Financial need and a desire to retain a sense of self-identity as a *shakaijin*, in order to not succumb to a 'disgusting' lifestyle of leisure that she half-disapproved of and half-longed for, combined to keep Sachiko-san working beyond her planned retirement.

Both these examples can be seen as indicative of a wider anxiety surrounding productivity in neoliberal societies in which the individual is split into both the 'entrepreneurial self and the self-exploited proletarian'.[41] Retirement offers the chance for people to live out desires and plans outside of the oppressive demands of work which are so common in Japan. However, in neoliberal societies the self becomes the oppressor, something that does not disappear when one retires. As sociologists Moore and Robinson argue,

> psychological changes arising from precarity contribute to the formation of anxious selves who have internalised the imperative to perform a two-part subjectification of workers as observing entrepreneurial subjects and observed, objectified labouring bodies.[42]

Kawase-san's sense of spending time productively was dependent on the routines he had established during his working years, without which he felt anxious and disappointed in himself. Similarly, Sachiko-san, as a self-defined 'typical Japanese worker', was unable to let go of the routines she had lived with throughout her life – yet her continuation of extreme commitment to work was also prompted by the precarity she faced as a single woman who had spent her working life in irregular employment. As the next section will show, women face a multitude of difficulties due to their participation in irregular employment, adding an additional layer of concern regarding decisions around retirement.

Irregular work

When Prime Minister Abe Shinzo entered office in 2012 and launched his 'womenomics' initiative, 3.5 million women joined the workforce. However, two-thirds of them are irregularly employed or part-time workers,[43] meaning that they often lack pension benefits, opportunities for advancements and job security.[44] The National Pension system (*kokumin nenkin*) is a mandatory program that all registered residents aged between 20 and 59 are obliged to contribute to. However, those in 'irregular work', which includes part-time, temporary and contract work, are beyond the scope of the national pension plan. In the same year that the 'womenomics' initiative was launched, 77 per cent of irregular workers were women, many from the older age cohorts.[45] Japanese studies researcher Helen Macnaughtan claims that the prominence of female workers in irregular positions directly correlates with 1960s employment models that foreground core male employment, supported by a temporary and cheap female workforce.[46] Many of my female participants found themselves engaged in such work after child-rearing and re-entering the workforce; others had been employed in such jobs throughout their lives. Today, despite record numbers of women participating in the labour force (71 per cent of the female population),[47] their financial future is not promising. Many lack savings, having suffered from one of the largest wage gaps among advanced economies.[48] In addition, policies that incentivise women to remain in part-time, low paying jobs in order for their spouse's salary to qualify for tax deductions have contributed to keeping women in irregular employment.[49]

Sachiko-san had worked as an irregular worker at the same large educational establishment for the past 20 years. She had made several close friends at the company over the years, with whom she would often meet up. One evening they met for *nomikai* (meaning 'drinking party' – a common practice in Japanese culture, often organised by colleagues and social groups) at a dumpling (*gyoza*) restaurant for which she had a coupon offer. Over a dinner, they got talking about work. The group[50] lamented how their long-term participation in irregular employment meant that they had no benefits such as a pension of their own. Most of these women were married and were able to depend on their husbands' income and pensions, while those who were divorced were now able to claim part of their husband's pension after a law was passed in 2007.[51] However, Sachiko-san was not married. She had raised her daughter as a single parent, and was therefore in an even more precarious situation than her friends. While irregular workers in Japan often experience increased isolation and loneliness,[52] I found that it was precisely the precarity of their labour that embedded this friendship group in the networks of support that came to mean so much to them.

The messaging application LINE had been central (*chushin*) to their friendship group before Sachiko-san had made the move to her parents' home in Fukushima. Before meeting, there would be a flurry of messages in their group chat, with stickers saying good morning and exchanges about their plans. Sending photos of her blooming garden via LINE to her friends, Sachiko-san shared moments of her new rural life. Although she could no longer go on regular *nomikai* dates with them in Kyoto, she maintained her friendships at a distance through her smartphone.

This digital connection was of some comfort, but it was not the same as seeing each other face to face. One of her friends remarked that every time she passed their usual hang out – a particular branch of McDonalds where they would meet for long chats over French fries – she experienced a pang of sadness. The *nomikai* dinners among this friendship group served as an opportunity not only to share enjoyable experiences, but to also support one another through the unfair labour practices in which they were embedded, for example by sharing the news that someone Sachiko-san knew from the Osaka branch of the company was starting a union to press for better conditions. With her move to a rural area, away from such support networks, the smartphone became even more important to Sachiko-san for maintaining a sense of solidarity with her colleagues and friends.

As I argue in Chapter 4, which highlights the vulnerability of older men in retirement as physical spaces of sociality disappear, women are better situated to benefit from emerging digital spaces of mutual support in later life. In the case of the female friendship group discussed above, it was precisely the webs of affective labour with which employment bound these women that made them hesitate to retire fully and kept them employed in low-wage, part-time jobs. The friendship groups that emerged through shared employment then transformed into sources of care and support as these women aged and approached retirement – and it was this affective labour that they prioritised as meaningful over the sense of fulfilment to be achieved through work.

Several members of the friendship group were employed as irregular workers at a catering company. Tomoko-san, aged 58, explained how she had found this job in her early fifties and had not yet thought about retirement. Because the work was quite physically demanding, the friends anticipated perhaps reducing their hours if they found that it was getting too much in the future. There was no company policy regarding an age limit for workers and many people of their age still worked there. They belonged to a LINE group with colleagues that was called '(company name) Family'. The women also enjoyed interacting casually with younger colleagues through the exchange of stickers which made them 'not even think about our age', as Tomoko-san explained. Despite the lack of financial incentives to remain employed, it was the social and personal benefits that these women most valued, now being facilitated through their smartphones. 'I finished raising kids and I had more time to myself so I wanted to work in some way. I began this job and my world changed,' explained Tomoko-san.

Through their employment, even if irregular, these women were choosing to engage in a different form of labour than raising their children. For them it felt like a fresh start, and it was difficult to imagine them wanting to retire soon. They framed their employment in terms of personal choice and a desire to stay busy, rather than a moral duty dictated by societal standards that associate employment with social worth.[53] They were engaged in projects of self-making that revolved around developing different identities than as mothers and wives. However, their notion of self was still defined in relation to practices of care. Through employment they became embedded in networks of care with their colleagues, many of whom became close friends. The women also saw working as a form of self-care, without which they might decline, as evidenced in Sachiko-san's concerns about retiring to the countryside.

The experiences of these women demonstrate a wider trend in Japan that perceives retirement not a time for leisure and rest but rather as a period of 'self-making',[54] something that I argue starts before retirement in the case of women to include taking up irregular employment after 'retiring' from the labour of raising their families. Notions of selfhood in the Japanese context have received much scholarly attention, with the individual positioned as inseparable from society.[55] Japanese concepts relating to the self within society will be explored in Chapter 4, but for now it is important to note, as anthropologist Emma Cook does, that there is no single 'Japanese self', rather 'a complex and often contradictory negotiation of selfhood'.[56] However, normative models of personhood do exist, and these are highly gendered. With stable employment central to predominant Japanese notions of masculinity on the one hand,[57] my ethnography points towards older women embracing irregular work so that they too can participate in *shakaijin* identity out of choice.

This practice of self-making through work is increasingly being mediated through digital technologies. The women working at the catering company, for example, now use an app to schedule their part-time work. They feel they have a greater degree of control over which shifts they accept, compared to the time before they had smartphones; then they would have had to speak to their employer on the phone and often found it difficult to say no. Being able to turn down shifts that they do not want is easier through the app, since the intense feeling of social proximity is removed. In this final section I show how ideas of what can constitute post-retirement labour are expanding, due to the rise of digital platforms that facilitate entrepreneurship. This expansion of opportunities has implications for the ways in which work in later life can be more successfully aligned with personal interests, rather than being dictated by the offerings of Silver Human Resource Centres or potentially exploitative employers.

'Post-retirement'

Social policy researcher Macnicol asks the pertinent question 'Why do people retire?'[58] He argues that retirement, while now widely accepted as an anticipated life stage and a phenomenon associated with welfare, is in fact a product of the last 130 years. In rural economies of the past, agricultural workers continued working for as long as possible. In my rural fieldsite of Tosa-chō today it is almost unheard-of to meet an older person not engaged in some form of small-scale agricultural labour. For example, Misako-san, aged 89, was always to be found in the vegetable garden outside her home. When she passed away in the year after I returned from fieldwork, I was told that instead of a long period of decline and hospitalisation she had worked in her garden on the day that she died. In rural Kōchi life after retirement is also largely dependent on financial security. While some people I met had previously worked in cities such as Osaka and Tokyo and then retired back to their hometown (known as 'U-turners', a phenomenon that will be discussed in Chapter 7), other people had never left the places where they were born. It was often those who had never left who had less disposable income and fewer choices about the kinds of post-retirement labour they could engage in. Many of these people had been wood workers – once one of the dominant trades of the town, but now no longer a profitable business because of cheaper imported wood coming from China. For women too there was pressure to maintain extended generations of families where men were unable to provide the required income.

Many of the people I met in both Kyoto and rural Kōchi were required to work well into old age because of financial need. Kioko-san, for example, is aged 88 and the owner of a local clothes shop in Tosa-chō. Small in stature, she has a lively sense of humour and is a formidable cockroach killer. The shop supports her entire family, including grand-children who live nearby. Despite being assisted by her daughter-in-law, the work kept Kioko-san busy from 6 a.m. until late into the evening; she also undertook tailoring on the side. She felt that the work kept her busy and active, and allowed her to see customers and friends every day which she valued. She did not have time to attend the local elderly club, which she felt was rather for people who had nothing to do. Despite this posi-tive attitude towards work, Kioko-san was conflicted: 'I heard that they are going to do a survey of happiness in this town. I am going to tell them that I am not happy,' she declared after talking about how she felt a lack of appreciation from her family for her continued work which sustained them during the unemployment of her son.

For people like Kioko-san, retirement exists only as a concept appli-cable to other people, not as something that might be a reality for them. There was little expectation among many of the older people I met in rural Kōchi and Kyoto that they would ever seek a complete withdrawal from various forms of labour. 'Jobless retirement' emerged in the twen-tieth century as a new kind of retirement due to a combination of the sectors that employed older workers shrinking and an increasing empha-sis on productivity.[59] Rather than approaching retirement as a state of joblessness, I found that it heralded a period of transition through which people changed jobs or engaged indefinitely in unpaid forms of labour. Yet there was also a sense that this was a time for exploring long-held interests and expanding hobbies and passions into entrepreneurial pur-suits. This topic will be further explored in Chapter 8, which consid-ers life purpose; there we will meet people who are seeking fulfilment and creative expression in later life. Since the baby boomer generation started to retire in the mid-2000s, terms such as 'second life' (*dai ni no jinsei*) became buzzwords in Japan.[60] However, I found that the deciding factor in whether people were able to benefit fully from opportunities for personal development in later life was typically financial security.

Keiko-san, who found work at the Silver Human Resource Centre as outlined above, explained that rich people can spend their retirements travelling and fulfilling their hobbies, but that she and Takashi-san could not. The couple's main hobby is visiting Japanese castles. They have vis-ited over 100 already and proudly possess a logbook with stamps from all the castles to which they have been. They do plan to visit more, but their leisure time in retirement is anticipated to be limited. Keiko-san thinks they probably will not be able to complete the stamp book in their lifetime (even though it appeared they already had stamps on most pages). However, they still hope to be able to enjoy a 'full moon' (rather than a honeymoon) holiday together at some point to celebrate this next stage of their lives, referencing an advertising campaign by Japan railways in the 1990s that offered a rail pass to middle-aged married couples.[61]

Keiko-san and Takashi-san, while grateful to have found post-retirement jobs, saw their work at the Silver Human Resource Centre as temporary; it was a stop-gap while their Airbnb business was closed during the Covid-19 pandemic. They chose to work at the Centre because they felt that otherwise they would have 'too much' free time. Rather than wait at home for the pandemic to end, a period of inevitably uncertain duration, they preferred to spend their time productively. However, their sense of productivity is not the sole motivator behind their labour choices. In wanting to re-start their Airbnb business, the couple chose to participate in an activity that connects them with people from all around the world and provides a personal sense of satisfaction.

I met many similar people in Kyoto who were in their sixties and were renting out rooms in their family homes to the many tourists who now flocked to the city. With empty bedrooms left by children who did not want to live together as a multigenerational unit, their parents were able to earn income through the platform and to enjoy the associated sociality. Several people I met in fact made very little profit from renting rooms on Airbnb, given the amount they spent on entertaining their guests, producing lavish meals and taking them sightseeing. For these hosts their participation in Airbnb was more of a personal hobby and a way to meet international people, legitimated as a business endeavour.

The Airbnb business was the responsibility of Keiko-san to manage. Out of the couple, she was the one more proficient in using the laptop and smartphone, skills that were necessary for the business to thrive. I found a similar pattern with several other couples of a similar age. One man in his late sixties from Kyoto would drive his wife to her computer class once per week; there she would learn how to use spreadsheets and other programs necessary for their Airbnb business. He would jokingly declare that she forgot everything as soon as he picked her up again, but it was clear that she was the one who possessed the digital skills in their household.

The rise of peer-to-peer digital platforms that enable individuals to participate in the sharing economy appealed particularly to my female research participants because of their affinity with new technologies and desire to develop meaningful work. However, feminist theorist Nancy Fraser warns of over-romanticising market-based solutions to women's poverty which promise greater choice and independence, such as micro-enterprise; in reality these often occur in the context of reduced services for the poor.[62]

Meanwhile the gender divided labour in the Katsura household (pp. 51–6) could be witnessed every evening during the time that I stayed with them. After dinner Takashi-san would lie on the floor, partially under the blanket spreading out from the low heated table (*kotatsu*),[63] while Keiko-san sat on the floor on the other side of the table, leaning against the sofa. From there she would half-watch the television while working on her laptop. The laptop was Keiko-san's domain. It was used for managing their Airbnb business and household accounts which since the First World War have been the responsibility of housewives in Japan.[64] Takashi-san joked that his wife had told him not to touch the laptop, to which Keiko-san acted shocked and gesturing towards it replied 'please, go ahead!'

In addition to running the administrative side of the Airbnb business, Keiko-san also makes breakfast for the guests. While domestic labour is typically the responsibility of women, during the time in which I stayed in the home of Keiko-san and Takashi-san I could see that while cooking was Keiko-san's domain, cleaning was Takashi-san's. If I ever tried to wash up my plates after dinner, he would shout over from the floor of the living room for me to leave my plates, saying simply 'my job!' Every night after Keiko-san had gone to bed, I would hear Takashi-san rolling a lint roller over the entire living-dining-kitchen floor to pick up stray cat hairs. He took his job seriously. Their participation in Airbnb blurred the private domestic sphere with the public sphere of business, and so the gendered labour roles associated with the domestic and public sphere also became blurred.

Rather than being a timeless national tradition, the 'family system' of a patriarchal unit promoted by the Japanese state, as discussed in Chapter 2, has been subject to change throughout the twentieth century.[65] The rise of the woman-centred family in the mid-twentieth century, with wives responsible for raising children, organising neighbourhoods, caring for older relatives and looking after the family finances, was both a result of women activists and the state.[66] Yet with women increasingly returning to the workforce after childrearing, as seen in this chapter, management of the household is often undertaken in addition to paid labour. This dual labour was a primary reason why many of my younger research participants were refusing to get married and start families in the first place. Those who did required their husbands to belong to a new 'type' of man, known as *ikumen*, who would actively participate in childrearing and other forms of domestic labour.[67] Though in reality the high-pressured employment conditions of the younger men I knew made it very difficult for them to do so despite their best efforts.

Conclusion

In one's sixties the notion of 'becoming a baby again' that Keiko-san referenced towards the beginning of this chapter has taken on new meaning. Traditionally referring to a time of increased dependence, 'becoming a baby again' in later life now takes on new meaning as older adults are having to learn new skills and adapt to new forms of labour – from men doing housework to people taking up new jobs through Silver Human Resource Centres. Just as a baby needs to learn how to be in the world, the adoption of new forms of labour in later life can be challenging, but are a central part of self-making.

This chapter has explored how gendered norms and the use of technology shape the everyday lives of older people through their continued participation in paid and unpaid labour after retirement. Building on Chapter 2, which demonstrated how the desire to reduce the sense of becoming a burden in old age is a mechanism for refusing the categorisation of 'elderly', I have argued that continuing labour in later life is also part of a wider morality around ageing. Such morality places responsibility on older adults to demonstrate care for others in various ways.

The ethnographic examples discussed in this chapter show how the perceived social and health benefits of continued work are believed to outweigh deficits such as low pay and irregular employment in people's decisions to continue working in later life. People value staying employed because of an association between retirement and physical, and even moral, decline. The chapter works with the *shakaijin* (member of society) identity that is formed through entering employment.[68] My research found that people, especially men, navigating the transitional period of retirement at the other end of their working lives do not want to leave behind this identity. The moral imperative to work can also be driven by filial duty, as in the case of continuing family businesses.

I have aimed to demonstrate how digital technologies including smartphones are starting to expand the range of opportunities for paid labour in later life. These in turn allow people to maintain a sense of continuity with their previous working identities, as well as heralding a turn towards more meaningful work – especially for women, who I found to be more engaged in learning new digital tools. The ageing population of Japan, and its corresponding implications for the labour force and economy, amplify the 'crisis of care'[69] that predominantly affects women across generations in ageing societies around the world, as they have to balance care duties with paid work. I have tried to highlight the distinctive situation in Japan in which older women often lack the same financial independence as men, contributing to a reluctance ever to fully retire. As digital skills continue to flourish among older women, it will be interesting to observe the repercussions on their labour prospects in later life and the degree to which it redresses established gendered labour roles.

The ethnographic material shows how labour choices are never free; there is no a clear-cut dichotomy between individual choice and societal burden. However, the smartphone and digital platforms are emerging as alternative sites for older people to conduct business and extend their participation in various forms of labour that are potentially more meaningful and under more of their control than was previously available to them. The smartphone is thus emerging as a site for the maintenance of *shakaijin* identity beyond the official retirement age, in parallel with official measures taken by the government to increase labour force participation among older people. More broadly, the uptake of digital technologies by older adults is facilitating the emergence of new forms of labour, and indeed new patterns of gendered labour for retirees.

Notes

1. Brinton and Oh 2019.
2. Crawford 2021.
3. Shakuto 2018.
4. Lamb 2014.
5. OECD 2019b.
6. Jones and Seitani 2019.
7. United Nations 2020.
8. Oshio et al. 2018.
9. Statistics Bureau of Japan 2022.
10. Oshio et al. 2018.
11. Kobayashi et al. 2022.
12. Chau 2021.
13. Japan Pension Service 2024.
14. Kobayashi et al. 2022.
15. Headquarters for Japan's Economic Revitalization 2019.
16. Oshio et al. 2018.
17. Coates 2019; Germer et al. 2014.
18. Borovoy and Ghodsee 2012.
19. Cook 2016.
20. Traphagan and Knight 2003.
21. Shakuto 2018; Mackie 2003; Nakatani 2006.
22. Mackie 2015,10.
23. Garon 2010.
24. Kano 2016, 4.

25. Kano 2016, 6.
26. Kano 2016, 9.
27. Fraser 2016.
28. Moore 2017.
29. Gagné 2021.
30. Roberson 1995; Cook 2018.
31. Gagné 2021, 252.
32. Gagné 2021; Cook 2018.
33. Moore 2017.
34. Traditionally people celebrating their 60th birthday would dress in a red vest and hat and be given red gifts. Red is symbolic of new life: *aka-chan* translates as 'baby', but also literally means 'red person'.
35. Shakuto 2018.
36. Weiss et al. 2005.
37. Watanabe 2020.
38. Watanabe 2020.
39. Shakuto 2018, 190.
40. Gaining comparative insight from the wider ASSA project, I was able to see both the distinctiveness of the Japanese case as well as its commonality with other fieldsites. The most extreme opposition was with the Irish fieldsite (Garvey and Miller 2021), where people in retirement generally made almost no reference to their previous work and were completely focused on non-work activities. A third possibility became clear in Sao Paulo, Brazil, a city known for its inhabitants' self-identification with work (Duque 2022). In that instance the emphasis was on continuity with the status and type of work people had previously been employed in, while – as the above examples have shown – in Japan the focus is more on remaining part of the workforce, even if the job itself is much lower in status.
41. Rolnik 2011, 48.
42. Moore and Robinson 2016, 2776.
43. Merler 2018.
44. Zhou 2020.
45. Macnaughtan 2015.
46. Macnaughtan 2015.
47. OECD 2020a.
48. Zhou 2020; OECD 2020b.
49. The gender employment gap is perpetuated by policies that incentivise women to earn less than ¥1.5 million ($14,066) a year, so that their spouses can receive tax deductions of ¥380,000 ($3,563). Katanuma 2020.
50. I have combined ethnographic evidence gathered from two female friendship groups who participated in very similar working conditions at two separate companies into one story for the purposes of clarity and narrative.
51. Alexy 2011.
52. Allison 2013.
53. Gagné 2021.
54. Shakuto 2018.
55. Bachnik 1992.
56. Cook 2016, 14.
57. Cook 2013.
58. Macnicol 2015, 1.
59. Macnicol 2015, 1.
60. Toyota 2006, as cited in Shakuto 2018.
61. White 1992.
62. Fraser 2011, 310.
63. A *kotatsu* is a heating device used in Japanese homes in winter. It consists of a low wooden table with a built-in electric heater and a heavy blanket underneath that traps the heat.
64. Komori 2007.
65. Garon 2010.
66. Garon 2010.
67. Mizukoshi et al. 2016.
68. Roberson 1995.
69. Fraser 2016.

Image by Laura Haapio-Kirk.

4
Social relations: sustaining mutual forms of care (*yui*)

It was in the sticky heat of mid-August when I first saw the Aikawa rice terraces in Tosa-chō. The green of the crop was of neon intensity, varying in hue as it cascaded from the top of the mountain to the river below. The organic shapes of the fields, conforming to mountainside rather than machine, traditionally required human hands for cultivation. In May the fields had been flooded with water, flowing from one farmer's field to the next and requiring everyone to do their part to maintain the water supply. Residents talked of the mountains turned to lakes, at night their mirror surfaces a dizzying upside-down world, shining with the light of thousands of stars. In the autumn the window of time to harvest a rice crop was so short that, before the development of machinery small enough to work single-family fields, planting and harvesting rice required neighbouring farmers to work one another's fields in a labour exchange system called '*yui*', which literally means 'to tie' or 'to join'.

The study of '*yui*', or labour exchange, has been at the heart of research into social relations in rural Japan during the twentieth century. Anthropological research first focused on the vertical relationships embedded in co-operative rural social relations between tenant and land-owner,[1] then progressed to research that demonstrated the way in which *yui* cements horizontal relationships between mutually co-operating small-scale farmers.[2] However, the term is not only of academic interest. A migrant from Tokyo to Tosa-chō explained that the presence of *yui* in this remote spot was at the core of her decision to relocate there with her young family. When describing the term, she mimicked the tying and twisting of a rope with her hands. This historical term still holds value and meaning; in both in rural and urban fieldsites people were engaged in forms of reciprocal care that enabled them to dwell comfortably within a wider context of insecurity.

A White Paper from the Japanese health ministry, released in 2023, called for the strengthening of digital forms of mutual support to tackle the increase in isolation and loneliness due to population decline and the pandemic.[3] However, at the time of my fieldwork, conducted between 2018 and 2019, it was primarily offline forms of reciprocal labour that emerged as central to the communities I worked with. This chapter thus focuses on examples from both rural Tosa-chō and urban Kyoto that show how the legacy of *yui* is expressed in a variety of ways in contemporary society, such as in the prevalence of volunteers (*borantia*), among older people. In so doing I build on the argument of Chapter 3, which explored the moral and social implications of paid and domestic labour in later life. I show how volunteering, in comparison, is driven by ideals of mutual assistance rooted in the legacy of *yui,* which I found to be equally strong in the urban setting as in the rural.

In demonstrating the importance of physical proximity for the practice of *yui,* I reveal the limits of digital platforms for mutual care. I argue that physical spaces of sociality continue to be important, especially for older men, yet discovered that such spaces are becoming limited in an era of rapid development in central Kyoto and of rural depopulation in Tosa-chō. I also show how physical proximity can lead to feelings of surveillance and over-closeness. Such emotions are then counteracted through digital practices that seek to reduce the social gaze while maintaining connection.

Civil society

Prior to post-war housing and farm reform, compulsory and hierarchical collective labour practices, known as *yui*, were commonplace in rural Japan. They involved communities in communal rice planting, roof thatching and other labour-intensive tasks, organised around the principle of powerful main (*honke*) families and smaller branch (*bunke*) stem families.[4] Today the term is part of official public discourse on volunteering, such as around disaster response, as seen on a government website called 'Tomodachi (friend) Disaster Prevention Edition'. A handwritten character (*kanji*) for *yui* adorns the website. It is made up of two halves – the left means 'thread' and the right means 'to firmly tighten'. The website emphasises the ability of people to join forces in times of need, like threads in a rope.[5]

Anthropologist Bridget Love draws our attention to the way that in contemporary Japan *yui* is appropriated along with other terms, for example 'community bonds' (*tsunagari*), 'connections' (*musubitsuki*) and 'mutual assistance' (*tasuke ai*), as nostalgic metaphors that dominate the language of social welfare planning. Love argues that in co-opting ideas of local renewal from a nostalgically remembered past, localised care in the form of neighbourhood civic action is drawn upon, rather than relying on limited government support in an era of decentralisation. However, rural towns such as Tosa-chō are particularly hard hit by societal ageing and depopulation, making such self-sufficiency increasingly difficult to sustain. These dual trends serve to drive an urgency around rural revitalisation through migration, to be discussed further in Chapter 7.

Local voluntary associations, such as tenant unions, resident, neighbourhood or ward associations, have a long history in Japan.[6] Their significance has grown with modernisation and the decline of the scope of kinship and local hierarchical structures for organising collective co-operation.[7] Rural areas of Japan have historically been the sites of early forms of civil society, as farmers and entrepreneurs formed alliances and networks of production, credit and sociality.[8] Volunteering, and civil society in general, gained widespread prominence in Japan after the 1995 Great Hanshin-Awaji (Kobe) Earthquake, specifically in preparedness, response and recovery.[9] In response to this disaster, which killed more than 6,400 people and left 320,000 homeless,[10] the rise of volunteerism was marked by a shift in how people saw their own responsibility as individuals within society; it empowered people towards self-governance and community solidarity, rather than relying on perhaps uncertain governmental assistance.[11] Anthropologist Inge Daniels, who has conducted ethnography in Japan since the 1990s, notes that a sense of individual responsibility has always been present, yet this shift in attitude was more about caring for others outside one's immediate family and community because of an awareness of the incapacity of the state.[12]

It was at this time that networked PCs emerged as instrumental in the organisation of mutual aid and activism, and in spreading an affective and emotional response to crisis situations.[13] With empathetic connection heightened by digital and traditional media networks, volunteering increased and become a highly visible phenomenon in Japan.[14] Sociality has been shown to be a key factor behind the motivation of volunteers,[15] who thus reveal both altruistic and egoistic reasons for participation in various forms of volunteer work.[16] Based on ethnographic research with volunteers, anthropologist Lynne Nakano argues that while volunteering is a personal choice, the emphasis on volunteer labour in elderly care reflects how responsibility is placed on local social networks of citizens to help one another, rather than the state, to alleviate family care.[17] The Covid-19 pandemic has provided further evidence from around the world of how crisis can spark local communities into action, with mutual aid groups rapidly responding to the needs of vulnerable populations to compensate for political and economic deficiencies.[18]

Volunteers who occupy leadership roles in rural communities are often key agents in addressing issues pertaining to rural life.[19] In Tosa-chō the people most often found in such roles were men in their late sixties and seventies; they were native to the area and were elected to leadership positions after working in other kinds of employment. Such age and gender profiles of village chiefs and community leaders are common in rural areas of Japan.[20] Older women tended to run community social groups, such as the elderly club and a club for new parents and older people to mix. Kawamura-san, a man in his early seventies who was running for local government, explained that at the core of his campaign were ideas for how the town can attract more people to move to the region. He is a member of a club with his friends whose purpose is to think about how to achieve rural rejuvenation. They meet regularly in each other's homes to discuss ideas, report on the concerns of local people and plan ways to put their ideas into action.

When I met Kawamura-san in the home of one of his friends, he was wearing his club's jacket with his nickname embroidered onto the sleeve. For him rural regeneration had become a passionate hobby that provided him and his friends with a social life and gave him a sense of purpose. Working towards the goal of sustaining, and hopefully increasing, his town's population was something that Kawamura-san recognised as urgent, even though many of the local people he spoke with did not have the same conviction. He explained that because the region has the facilities and infrastructure for a population double its actual size, some people do not feel that the issue of depopulation affects them directly. The difficulty of attracting migrants was compounded with the problem that many did not become permanent residents:

> We are trying to attract migrants, but the problem is that there is not much work here. But people can create their own work if they have ideas. They can do small-scale farming, and people from cities are quite good at setting up networks of sharing in this community so you can get by even if you do not earn much money. We can try to attract people to live here, but there is no guarantee that they will have babies.

Kawamura-san noted that there were a lot of single people in the area – men and women who were middle-aged and over but who had never married or had families. He put this down to life being very comfortable these days as a single person, with people using their smartphones to stay connected to friends and family instead of having to live with them. In this way, Kawamura-san saw the smartphone as contributing directly to the 'convenience' or viability of single life – and by extension to the rise of the ageing population in the region and the country as a whole. However, he observed that it will be these single people who face the biggest challenges in ageing, as they lack the proximate support of family when they become frail or ill. At the same time, the smartphone is central to these 'networks of sharing' that he identifies as integral for migrants to be able to settle here successfully.

Ethnographies focusing on older rural Japanese dwellers have shown them to be 'caretakers of collective wellbeing', for example through the management of religious rituals.[21] Young rural residents have been shown to be shaping their own versions of modernity in the countryside.[22] Conspicuously absent from this generational picture are the middle-aged (those in their fifties and sixties). In Tosa-chō they are the missing generation, comprised of those who left for employment in the city and never moved back. So while older residents take on the roles of community group leaders, some wonder who will replace them in the coming years, especially when they may begin to exhibit signs of cognitive decline. I observed that younger migrants and locals are often involved in non-institutional webs of support and care with older residents, outside of the structures of community groups and based more on locality and friendship.

Japanese society

In the 1970s anthropologist Nakane Chie famously argued that social relations at all levels of Japanese society, such as between boss and company worker, within kinship systems and within communities, are founded on the vertical structure of the rural landowner and tenant relationship, producing a 'vertical society' (tateshakai).[23] Nakane's book *Japanese Society* has been critiqued for presenting a homogenising vision of social relations in Japan,[24] and the wider genre of *nihonjinron* popular and academic literature, of which her book is part, is generally regarded by anthropologists as essentialising and mystifying Japanese culture.[25] Conversely, the anthropologist Tatsuro Suehara argues that more recent studies of rural social relations demonstrate the importance of horizontal relationships of mutual assistance for maintaining communities. Such studies explain the possible discrepancy with previous work as being to do with the time that the research was conducted. Earlier research conducted from the 1930s to the 1960s presents a very different view of rural life to that conducted in the 1970s and later, when rural Japan underwent massive changes.[26] Indeed, Suehara notes that since the 1990s the meaning of *yui* has expanded from its original agricultural roots to encompass wider forms of mutual care; it is now often used in the naming of community and volunteer organisations.

Nihonjinron literature typically compares Japan with 'the West' along a variety of ahistorical and homogenising diads, such as 'individualism/groupism', 'independence/dependence' and 'egalitarian horizontal ties/hierarchical vertical ties'.[27] In this chapter I focus on 'horizontal relationships', i.e. those between peers and generally non-kin – not to engage in reductive analysis of social relationships, nor because of the scholarly legacy of *yui* within anthropology, but because this proved a key concern in the context of a perceived decline in such relationships, both in my urban and rural fieldsites. Rather than focus on a reified idea of 'groupism', this chapter's focus on *yui* is more in line with scholarship on 'contextualism' or 'interpersonalism' (*kanjinshugi*), highlighting the importance of 'mutual dependence' in interpersonal relationships in Japan.[28]

Local people in Tosa-chō explained how their small communities of hamlets and villages were tied together through generations of mutual assistance. However, machinery assists with the labour today, and there are fewer people practising rice farming. Next to the Aikawa terraces, introduced at the beginning of this chapter, in a particularly scenic spot the local council had installed a speaker atop a large pole, designed to play an old harvesting song at the press of a button. This song would have been sung historically by communities as they worked together. In this remote region, at the top of a winding mountain road, today it is tourists who hear this song at the press of a button, a vestige of an idealised form of rural sociality.

The people who remember these songs[29] are now few and elderly, while abandoned rice fields with their Van Gough swirls of overgrown rice mixed with wild vegetation are a common sight in this depopulating area. The association of mechanised farming practices with a reduction in co-operative sociality exists in parallel with a common fear expressed by many older men in both Tosa-chō and Kyoto: that the smartphone is accelerating a movement towards a more individualised society, increasing the distance between people and making single life 'too convenient'. I often heard variations of 'It's not human!', lamenting a perceived loss of face-to-face communication with the rise of the smartphone. Meanwhile older women tended to be far more positive about adopting the smartphone, proudly showing photos of grandchildren that had been sent to them via LINE. This gendered discrepancy towards the same device was one of the first things that struck me during fieldwork. It has continued to be a central thread throughout the research, discussed particularly in Chapters 2 and 5 of this book, but woven into the text throughout.

The negative appraisal of the smartphone fits with wider narratives of disconnection that have been gaining momentum in Japan since the early 2000s. Terms such as *muen shakai* ('relationless society') are common in popular discourse about the current state of Japanese social relations, which are challenged by increasing solitary living.[30] In this context, digital technologies are just one part of a wider set of shifts seen to be negatively affecting society, especially its older people. In 2010 NHK (Japan's national broadcasting agency) aired a documentary about the precarious state of Japan's ageing society, the title of which, '*muen shakai*' (relationless society), was subsequently widely discussed in the national media (Nozawa 2015). The term has now been adopted into everyday life, capturing the zeitgeist: people felt they were living in an era of individualisation and isolation that led to such phenomena as solitary deaths (*kodokushi*) and unreported deaths among the elderly.[31]

Paradoxically, after the triple disaster of earthquake, tsunami and nuclear meltdown that occurred the following year on 11 March and the subsequent rise in civil society, the year 2012 was proclaimed as 'the era of human and personal relationships' (*en no jidai*).[32] Such a turnaround demonstrates how zeitgeists are, by definition, contextual and subject to change. In anthropologist Anne Allison's work *Precarious Japan* an anthropological gaze is cast on the sense of unease and anxiety that permeates life for those who experience 'an affective turn to desociality that, for many, feels painfully bad'.[33] This bad feeling is precisely what may prompt some people to engage in purposefully social activities such as volunteering. However, the issue of social isolation remains prominent in Japan.

During the 1990s and early 2000s social isolation was typically considered a youth issue, affecting school dropouts and those unable to fit into society.[34] However, in 2015 a Cabinet Office survey suggested that now more *hikikomori* (social recluses) are aged between 40 and 64: an estimated 613,000 individuals, compared with the 541,000 who are aged between 15 and 39. The majority of this older cohort (75 per cent) are men.[35] With the rise of personal digital technologies, the phenomenon of *hikikomori* has been linked with internet and smartphone addiction.[36] My neighbour in Kyoto was a social recluse in his thirties or forties. I would hear him playing computer games at night, but only ever saw him briefly at the door when a woman delivered groceries once a week.

A doctor in Kyoto described how in her small practice on the outskirts of the city they had five registered *hikikomori* patients. The doctors would visit them at home, at the request of worried family members, and speak with them via closed bedroom doors. Her opinion was that the issue often stemmed from disrupted family relationships and thus she advocated for a medical approach which took into account the health of the entire family as a unit. While *hikikomori* represents an extreme form of social disconnection, I found that many people shared a sense of social relations being more generally in a state of jeopardy.

However, my ethnography also complicates the picture. In bringing out the gendered nature of the issue and looking at how social distance can be desired in contrast with over-burdened traditional forms of sociality, my findings are in line with work by anthropologist Iza Kavedžija who studied elderly attendees of a 'community salon' in Osaka.[37] Kavedžija found that people appreciated engaging with others in the salon in friendly but not intimate ways; they did not wish to return to previous forms of neighbourhood sociality that were perceived as too close and overwhelming.[38] This kind of balancing act was highly evident from my observations of how people maintained their privacy in Kyoto, only socialising with neighbours during socially sanctioned periods such as festivals, as the next section will show.

Neighbourhood sociality

The central Kyoto neighbourhood in which I lived for most of my fieldwork, which I have termed Kitano, was a place where people tended to keep themselves to themselves. Walking down the street, neighbours would greet each other and would remember to thank each other for gifts previously received (people explained it was polite to thank someone three times according to Kyoto etiquette). They might enquire about each other's grandchildren in a polite and friendly way, but apart from that neighbours maintained very private lives. For those living in apartment blocks (*manshons*), it was felt that this sort of living further reduced contact with neighbours compared with living in houses, where at least one might greet a neighbour on the street. Neighbourhood sociality is largely institutional, for example through neighbourhood associations – which, at around 300,000 across Japan, represent the biggest civil society organisation in the country.[39] In anthropologist Eyal Ben-Ari's ethnography of two suburban settings close to Kyoto, he found that it is predominantly women who manage such local neighbourhood associations, along with other community groups, despite their additional participation in paid labour.[40] This gendered labour aligns with the state sanctioned view of women as responsible for communities, as outlined in Chapter 3. Anthropologist Joy Henry's ethnography has also demonstrated the importance of such community systems in rural Japan.[41]

The 'neighbourhood chief' (chōnai kaichō) in Kitano would monitor neighbourhood issues such as recycling and was called upon to visit residents in times of trouble, for example when typhoons or earthquakes caused damage to properties. They kept an eye on frail elderly residents and generally maintained the smooth running of the neighbourhood. However, people would also jokingly complain that someone would only become a neighbourhood chief because they wanted to know everyone else's business. A degree of neighbourhood scrutiny was evident from the way in which my landlord asked me to give him my recycling so that he could put it out on the street on the appropriate day of the week; he was worried that his neighbours would see a foreigner putting out the recycling and would assume that I was sorting it incorrectly. To avoid any tension, my landlord preferred to put my recycling outside himself. Mostly these neighbourhood forms of sociality that maintained harmony were invisible, and in general neighbours respected each other's privacy and kept their distance.

A notable exception occurred at certain times of the year, when community festivals provided approved opportunities to socialise. For example, during the Gion Matsuri in July (one of the largest festivals in Japan), in which a large parade makes its way through main streets of Kyoto and smaller streets are full of stalls, private spaces such as driveways become semi-public venues in which neighbours may socialise with each other. In the festival that I experienced, my landlords and a couple of their neighbours sat at folding tables that they had set up in their driveway where their car was usually parked. Inviting me to join them, they bought me food from the pop-up stall of a local restaurant. A neighbour served home-made sweet shiso leaf juice mixed with alcoholic shōchū and we talked about their neighbour's forthcoming travels. Such public socialising was a very rare occurrence, but the festival served to blur the boundary between public and private. While my landlords may not have ever invited these neighbours into their home to socialise, setting up a table in their driveway invited sociality in an informal, semi-public setting.

One of the most social local community events I participated in was *Jizobon*. This Buddhist festival was held in August to celebrate *Ojizō-sama*, a bodhisattva known as the protector of children. On the morning of *Jizobon* I awoke to the very unusual sound of voices on the street below my building. At 7 a.m. neighbours were already gathering to put tables next to the small shrine built into the street-facing wall of my building that housed statues of *Ojizō-sama*, like many others in Kyoto. On these tables they placed red tablecloths, to which people added their offerings of food and sake. The tables soon became full.

Later in the morning of the *Jizobon* festival, neighbours gathered around to watch as a monk arrived and chanted sutras. After the ceremony the festivities lasted for a few more hours, during which the children ate shaved ice flavoured with matcha tea syrup and the adults drank sake and chatted. About 30 adults and children attended. So as not to take up too much space on the narrow road, which was still open to traffic, the games and socialising were hosted in the open garage of one neighbour, and in the entranceway to the home of another. Neighbours who saw each other only occasionally asked how each other's families were doing and people's grandchildren came from different areas to join in the festivities with their parents. As one resident told me:

> There used to be many children living here, but not anymore. Now we invite our children and grandchildren to come and join us for the festival. Even the priest is now old!

These moments of celebration provided stark reminders of the ageing of the local community. Young families were reluctant to remain living in the neighbourhood and take on the responsibility for homes that were themselves ageing and needing maintenance. It was common to see empty plots suddenly appear in the neighbourhood, the old buildings having been rapidly demolished ready for development. The photograph below shows a site in Kitano due to be developed into a 28-room hotel (Fig. 4.1).

During my fieldwork there were constant physical reminders of how the population demographics were shifting towards an ageing and shrinking population, as well as of the booming tourism industry in Kyoto. Old *machiya*-style wooden houses were transformed into hotels, while multi-storey apartment buildings designed for singles or small

Figure 4.1 A photo of an empty plot in central Kyoto. Image by Laura Haapio-Kirk.

families seemingly sprang up in a matter of months. Long-term residents said they felt that central Kyoto was changing beyond recognition. Saori-san, a woman in her mid-sixties who had grown up in the area next to where I lived, explained that the local *Ojizō-sama* shrine, which would normally have been the focus of their *Jizobon* festivities, had been demolished when the apartment building next to her house had been built. Since then the community had taken it in turns to look after the statues that normally lived inside the alter. This year it had been her turn to look after the statues and to host the ceremonies in her own home. She had felt anxious about the responsibility placed upon her to act as host, and was relieved when the festivities were successfully completed.

Saori-san's home, like many others I visited in Japan, had only a small area where it was possible to receive guests. Along the walls of her living room were display cabinets and bookcases with many of her late husband's collections of objects, among them large model ships. There was not a lot of space for a community to gather. One wonders about the future of festivals such as *Jizobon* where local communities can socialise, as these require outside space in a context where such areas are disappearing.

Facebook became an important meeting point to share sadness and disappointment as community shrines disappeared. The monk at my local temple shared a Facebook post with photos of a 'shrine memorial ceremony' he was called on to perform by a neighbouring community before their *Ojizō-sama* shrine was demolished. In the post he noted that the demolition was not only because of the location of the shrine; it was also due to fewer and fewer people looking after the statues. The neighbourhood association was relieved to hear that this monk's temple would care for the statues that had been passed down through generations. Comments on the post were mainly from older adults, expressing their dismay at what they saw as a widespread issue associated with the country's ageing population. They felt sad that younger generations were not looking after these local historical, religious and cultural sites.

Social media posts, such as by this particular monk, offer a space in which communities can express their feelings and gather collective momentum for change. The stark photographs of the demolition of the shrine that once protected the children of a community, and the subsequently empty lot, proved an emotive sight, sparking conversation and calls to action. We can see how in particular digital visual media via smartphones can provide 'new spaces for networked, *effective* civic responses and *affective* interpersonal responses', as put by anthropologists Hjorth and Kim.[42]

Many people would talk of the disappearing community, or in their words 'horizontal relationships' (*yoko no kankei*), as directly related to the changing design of housing; from another perspective, however, the changes may be seen as propelled by different desires regarding living arrangements. It was common to hear people say things such as 'a long time ago, there were no apartments, right?' when explaining the loss of community now felt in central Kyoto. I suggest that this emphasis on material culture as deterministic of changes to sociality has now morphed into a critique of the smartphone – widely seen as contributing to the convenience of single apartment living, implicating reduced need for physical proximity with others. In central Kyoto the rise in tourism has accelerated the changes to the community. Here I present the words of Nomura-san, a local resident and restaurant owner in his early eighties, who summarised the changes he has witnessed over the past 30 years.

Nomura-san: Until then [30 years ago] there were no people walking around my house, around this area. It was very local. A local area in Kyoto. There were no tourists, it was all traditional wooden houses [*machiya*]! Just residential. But you know, this apartment block [*manshon*] in front of us, maybe half [of the apartments] are occupied. The others just use it as a holiday home.

Laura: So what does that do to the community?

Nomura-san: Well before, of course, the community would look after all the kids in the neighbourhood. But now that has changed a lot. Since the area has become a different place. Also many buildings are no longer homes, they are apartments. So that makes it difficult for people to have a sense of community. *Jizobon* doesn't happen around here anymore.

Nomura-san identifies the problem of community disintegration as stemming from multiple parallel processes. First, the reluctance of younger generations to return to their ancestral homes means that there is no longer the kind of ancestral continuity that Kyoto people pride themselves on. As the eldest son, Nomura-san always maintained a close relationship with his parents in Kyoto, even during his working years in Tokyo, when he would visit them several times per year. Upon his return to his family home, he took over the family restaurant and his sons now also work in the kitchen. They also have their own separate ambitions, however, and it is unclear whether the restaurant will continue once Nomura-san stops working. Second, the rise of tourism has radically changed the nature of neighbourhoods in central Kyoto; apartments are rented out, with Airbnbs and hotels appearing on every street. Third, with the rise of apartment living, buildings no longer fit the idealised notion of what a 'home' should be, rooted in the traditional concept of multigenerational living.

The dissolution of his local community is, for Nomura-san, evidenced in the way that festivals such as *Jizobon* no longer take place around where he lives. As observed above, this festival was indeed at the heart of sociality in my neighbourhood, just a 10-minute walk away, although there was a growing sense among residents that the festivities were a vestige of former iterations. The precarious situation facing *Jizobon* festivals is in direct contrast to the large and well-attended *Gion Matsuri* festival, one of the biggest tourist highlights of the year for Kyoto. Many local older people I spoke to chose to avoid the huge crowds and oppressive July heat of *Gion Matsuri*; they preferred to watch the parade on television or from an air-conditioned building. The foregrounding of tourism over local communities was evident in how many temples, especially the larger ones, operated in Kyoto. However, the abundance of temples makes them well-situated to provide a locus of community support. While homes may be changing in central Kyoto, temples offer more permanence; through volunteer action, they can be re-appropriated for community use once again, as the next section will show.

Not leaving anyone behind

While very few people I met would describe themselves as religious, nearly everyone in Kyoto and Tosa-chō visited temples and shrines for annual festivities as well seasonal activities such as autumn leaf viewing. In a more everyday sense, Buddhist temples were important as sites of local community, especially for older people. The local temple on my street in Kyoto had several activities in which older people were engaged, including a choir, a Hawaiian dance group and a Boy Scouts network for former members. One of the monks was committed to finding ways to engage people in activities beyond religious festivals, including organising exhibitions, talks (one of which I was asked to give) and workshops, which then moved online during the Covid-19 pandemic. Such workshops, activities and community events created a backdrop of organised sociality that engaged members of the local community such as Haruki-san, in his mid-eighties, whom we met in Chapter 2. While he would never go to the activities organised especially for the elderly at the local community centre, believing that they would 'suck his youth away', Haruki-san was a constant presence at the temple, as we have previously seen.

Membership of the temple was declining, with an ageing and shrinking congregation of residents. Buddhist temples have traditionally been financially supported by local households in what is known as the *danka* system (*danka seido*). Today fewer people practise Buddhism; fewer people also live in the area surrounding my local temple, given how radically central Kyoto has changed. One of the monks explained that *Obon*, the festival for ancestors in August, was their busiest period. Then he would visit about 100 houses, performing prayers at older people's homes, but outside of this period they needed to find new income streams if they were to keep the temple going. The wife of the youngest monk explained how she was also getting involved in diversifying their income. She was organising activities aimed at tourists such as flower arranging (*ikebana*) and kimono lessons, which they would advertise on Airbnb's 'experiences' platform. While the temple expanded its reach to benefit from the rise of tourism in the area, especially in anticipation of the 2020 Olympic Games (held in 2021 because of the pandemic), they were also aware that pivoting away from their traditional model could put them at risk, especially if the numbers of tourists dropped.

This concern was proved astute in the two years following my fieldwork, in which the Covid-19 pandemic obliterated tourism in

Japan and around the world. I heard that as a result many hotels in central Kyoto closed permanently. At the time of writing, it remains to be seen what will happen to the city centre now that it has started to re-open to tourism again. One thing is clear: the local older adults who rely on temples as sites of sociality are having to compete with the reality of temples needing to think beyond serving local communities to survive.

Some residents recognised this disconnect even before the pandemic and during my fieldwork were seeking to find a solution that would serve the community. The Kitano ward office held a meeting for residents to discuss ways to improve life for local people. At this meeting Miyako-san, a resident in her early sixties, met several other women keen to start a community group to promote neighbourhood sociality in their area. One of the other women was married to a priest at a different local temple. She suggested that they work to make temples key sites of sociality and mutual support, much as they had been in the past. The women set up a LINE group to facilitate their discussions. Within one month they had developed their ideas sufficiently to apply for modest funding to cover their expenses. Miyako-san and her community group were motivated by a feeling of dissatisfaction with contemporary life; city life in particular, she explained, is more 'convenient' and therefore there is no longer the need to co-operate to survive. 'You don't have to be connected to a person to live any more,' she observed, something that she felt led to a decline in people's awareness of others and in the webs of community support that used to be commonplace.

The group felt that there were many people in society who were in danger from social isolation (koritsu). Those at risk spanned all generations, from elderly people to new mothers, recent retirees to children who might be finding school difficult. The women's application for funding positioned the group as mediators between temples and local residents; it also outlined their plans to establish activities that would bring multiple generations and different kinds of people together in one place, describing the mix of people they hoped to attract as being like a fruit bowl (gocha-maze). They took inspiration from previous activities that temples used to engage in within the community, such as after-school 'cram schools' (terako-ya), at which older adults could help students with their homework. The aims of this volunteer group were deeply aligned with notions of yui, believing that in bringing multiple generations together they might restore a cyclical notion of mutual assistance to the community (ii jyunnkan ga, komunitii ni dekiru). The idea was to stimulate autonomy (*jichi*) at the same time as supporting each other (*minna de sasaeau koto ga dekiru*), encapsulated by their slogan 'Someone else's problem, my problem' (*hitogoto jibungoto*).

Proximity was central to this volunteer group's plans to revitalise communities through mutual assistance. One of the main attractions of using temples as the locations for these activities is their proliferation within Kyoto. There are over 1,600 Buddhist temples dotted throughout the city, meaning that every resident is within walking distance of one. In the context of concern about 'connection' (*en*) within Japanese society, if kinship is no longer a guaranteed mode of sociality,[43] then perhaps connection based on locality emerges to fill the gap. Anthropologist Nozawa Shunsuke contends that solitude and social indifference serve to produce a fantasy of sociality, which he describes as the 'fantasy of the phatic', encapsulated in Japan by the idiom of *fureai* (touching together).[44] The 'fruit bowl' metaphor employed by the community group described above conveys a sense of everyone cosily mixed together, in contact yet still maintaining their own individuality. I will return to the notion of phatic connection in Chapter 5 which discusses visual digital communication, showing how it enables a sense of 'touching together' while also maintaining the benefits of distance.

The notion of warm connection also appears in Chapter 6 in relation to welfare initiatives. In the case of the public discourses of concern around the phenomena of solitary deaths (*kodokushi*), Nozawa argues that while people desire connection, their reality is often far more grounded in resignation which produces a 'fantasy of relationality' that prompts new forms of co-presence with others.[45] Such a fantasy, he argues, is what contributes to the prevalence of *ore ore sagi* ('it's me, it's me) phone scams, in which older adults are contacted by strangers who pretend to be relatives asking for money. His research participants admit that they would probably fall for such scams, simply because of wanting to give out of a desire to be connected.

Kinship relations based on obligation through blood are diminished in a depopulating society, with ties between strangers through choice (*kizuna*) emerging as central to living well. I found that the proximity of local communities, while nostalgically longed for, could also be associated with over-closeness and scrutiny. It is in this context that people are developing ways to facilitate connection based on proximity, while recognising the value of distance. I argue that digital technologies are an extension of these practices, allowing people to engage in 'tactical invisibility' to reduce a public gaze while also facilitating connection.

Alternative spaces of sociality

Kazuko-san, a stylish woman in her early seventies, lives in an old house in Kitano that previously belonged to her mother. After returning to Kyoto upon her mother's death, she decided to open the ground floor of the house as a semi-public space in which to hold events for the local community. She wanted to meet people after living away from Kyoto for many years, so she put a sign on the door advertising her first event – a wine tasting evening where people could also bring their own dishes to share. She collaborated with the local wine shop on her street, asking them to bring a selection of wines to try, and charged a minimal entrance fee to cover costs. This advert on the door attracted curious local people, such as Nomura-san, the restaurant owner described above, and various other people of a similar age. Her newspaper advert was also successful in attracting a couple in their early sixties who had moved to Kyoto from a rural part of another prefecture in order to participate in cultural events. However, most attendees came alone, and at the subsequent monthly events that I attended there were typically more men than women.

Kazuko-san listed her mother's home as an events space on Google Maps. She also formed a Facebook group which she would use to create event invites, in addition to compiling a list of email addresses which she would also use to remind people of upcoming events. Throughout my time in Kyoto I attended regular dinner and wine evenings hosted by Kazuko-san in her home, and got to know the small community (roughly around 20 people) mostly in their sixties and above, who attended on a regular basis. However, at any one event there might be several new people and the mix of people continually changed. A typical evening at one of Kazuko-san's events would involve attendees arriving on time, placing their home-made or shop-bought dish on a table and then taking up seats placed around the edge of the main room. Once all attendees were settled, a round of self-introductions would commence. Each person took it in turns to stand up and introduce themselves, sharing their occupations and where they lived in Kyoto.

The semi-formality of a public event, replete with self-introductions, allowed people to engage in social relations that were at once friendly but reserved. Before the wine tasting and eating would commence there were often scheduled performances. On one evening I gave a short presentation about my research, for example; on another someone who was learning *rakugo* (comedic traditional Japanese stand-up) gave a performance and shortly before Christmas we had a quiz. Over the 18 months of fieldwork and thereafter via Facebook and Instagram posts I witnessed friendships blossoming among the original dinner group members. Several female members often posted photos on Facebook, taken together at Kyoto temples or festivals they had attended.

One of the regular attendees, a quiet man in his late seventies called Sakamoto-san, saw the advertisement for the dinner group in the newspaper. He decided to attend after realising the need for sociality more keenly in later life. After pouring all his energy into his work, he found that in retirement he had few friends, while his wife had a social life that did not include him. Sakamoto-san was happy to enter retirement at the age of 60, having never taken holidays apart from a few days during 'Golden Week' in May, during which most businesses shut. Over 10 years later he fills his days with attending public lectures at Kyoto University and learning the piano, though at night he regularly dreams that he is still at work. 'I'm supposed to have quit by now! There must be something psychologically ingrained in me about work,' he reflects. Such dreams leave Sakamoto-san feeling slightly disconcerted, but he has many great memories from work that he still cherishes. As he got older, he realised how difficult it was to develop friendships, especially for men. He observed how they have often not had the opportunity to develop relationships outside of work prior to retirement.

As we drank green tea, sat on the tatami mats of his beautiful home in northern Kyoto, Sakamoto-san explained:

> I think women have more friends ... they have more opportunity to make friends. Because for men, it's all strapped up in work. There's no time for friends. You know, there is no time to meet people outside of work. Because I'm crazy about work and I think about it all the time. But somehow ... there's a hole in my heart. There's something empty. So, when I was young I wanted to accomplish. I was ambitious and I wanted a challenge. But when you get older, that ambition is umm ... it's harder. I think that's because you can't just think about yourself when you are older. Meaning, you have to think of yourself in a group. When you are younger, you just think of yourself, it's fine. But I don't think that's the case when you are older.

Sakamoto-san called me one day (from his landline since he did not have a mobile) to invite me to a lunch at his home with some of the other dinner group members. We enjoyed a relaxed afternoon together while his wife was at her 'Chinese culture group' (actually a mahjong club, he said with a laugh). Inviting group members to his home and serving his 'men's cooking' (*otokomeshi*), as he called it, was a way in which Sakamoto-san strove to build relationships outside of his immediate family and to put into practice his notion that 'you have to think of yourself in a group'. Like the 'fruit bowl' (*gocha-maze*) analogy as used by the temple volunteers, Sakamoto-san's understanding of later life as requiring more purposeful socialising was premised on the ability to be physically with his new acquaintances.

For Sakamoto-san the physical space of the home, both his own and Kazuko-san's, was a central site of sociality. For the women of the dinner group, however, their social relations expanded into online spaces, enabling them to participate in repeat get-togethers and online exchanges. Sakamoto-san's assessment that women have more opportunities to create friendships seems to apply not only to apply to the context of having the space and time to do so outside of work, but also to the fact that women tended to be far more open to maintaining friendships via the smartphone. Technology for them facilitated social proximity at a distance, as well as encouraging the exchange of visual digital phatic communication,[46] which I will examine in more detail in Chapter 5.

Of course, this was not the case for all older men. Miyagawa-san, who appears in the film below (Fig. 4.2), was a member of the former Boy Scouts' network at my local temple, described above. In his spare time his main hobby was the construction of model ships. This hobby, and the subsequent sharing of it on Instagram, became particularly important to

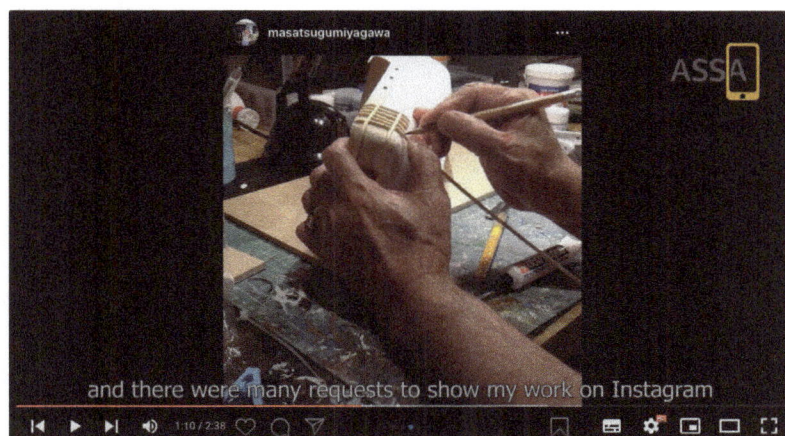

Figure 4.2 A short video about Miyagawa-san's hobby of model ship building which he shares on Instagram. https://youtu.be/nI3ZeR1n48Q.

him after he was diagnosed with cancer and during his recovery period. During this relatively isolated time he was able to receive encouragement from people on Instagram. As he states in the film, he believes that both his hobby and staying social, both online and offline, is now what keeps him healthy.

The value of distance

With a rapidly falling birth rate in Japan, forms of family care that used to take place between generations within one household are now increasingly dispersed among more distant kin and non-kin. Such forms of care often also take place outside of the home, and now increasingly online. I met many people in their thirties and forties who did not have children; several research participants in their fifties, sixties and older were also childless and often single. While they were often living highly independent lives and valued their autonomy, they were also enmeshed within various forms of care and support with family, friends and service providers. I found that in such cases the smartphone played an import mediator, facilitating the care given and reducing the sense of burden or trouble (*meiwaku*) that they felt they might be causing to those around them.[47]

For example, 93-year-old Sato-san was a flower arranging (*ikebana*) teacher in Kyoto who did not have children and relied on her niece for regular assistance, for example to change the water in her flower displays at various businesses and shrines. She valued her smartphone for the information she could access through it which facilitated her everyday life – from bus timetables to the weather. She particularly liked how it enabled her to do things by herself rather than rely on family. This independent approach accorded with Sato-san's general attitude towards life of embracing new experiences and forging her own path. She was able to message her local Seven Eleven store to order food, and they would send her photographs of tofu and pickles to check that these were the type she wanted. Likewise, when she heard something on television that she wanted to know more about, Sato-san would now look it up on Google instead of phoning her niece. In this way, the smartphone can be seen to facilitate a continuation of the independence and curiosity she valued so much throughout her life. As she stated:

> I've had that since I was very young. So curious, such a curious mind! I would always ask 'why?' And every time I did, they would answer 'because it is so!' [laughs]. If I have a question about

whatever is on TV, I would ask my niece. And she would tell me she will look it up later, but it's so much easier to just look it up on my phone myself.

Sato-san is supported in living independently by a web of family relations, as well as companies such as her local supermarket. Her smartphone is now central to how she communicates with both. The smartphone had also become important to her for another reason: it allowed her to keep learning the English language. She had taken classes in her seventies, but quit because she felt out of step with the students and was worried that she was becoming a burden on their learning. Since she got her smartphone she uses it daily to learn new English words as part of her brain training regime. However, an implication of Sato-san's framing of her smartphone use as something to relieve feelings of burden shows that there may be pressure on older adults to 'prove' their independence and capability through digital technology adoption. This did not seem to be the case for people of Sato-san's age; she was proud to be an early adopter of digital technologies and felt frustrated by her contemporaries who resisted even acquiring feature phones (*garakei*). However, it may be the case for the generation below her.

In stating that she no longer had to bother her niece with queries, might not Sato-san also be more vulnerable to misinformation and scams that specifically target older people, without the sounding board of a trusted family member? For others who may not be as socially embedded as Sato-san, with regular face-to-face interaction in addition to digital connection, the risks may be higher. Rather than framing the adoption of smartphones as facilitating independence, as we shall see in the next example, my research demonstrated how the device is emerging as central to the mediation of interdependence.

Maiko-san, aged 67, is one of Sato-san's students.[48] She is enmeshed in multiple webs of support, both online and offline (explored further in the co-produced comic preceding Chapter 9). She attends classes in *ikebana* every two weeks at Sato-san's home. Maiko-san lives alone and has no children. She teaches music and enjoys listening to live jazz in the evenings in Kyoto bars and hotels. Maiko-san is highly attuned to ways in which she might stave off boredom, something that she feels is one of the major threats to happiness as one ages. She often needs to attend medical appointments at the hospital and relies on her brother to drive her there. These trips are also opportunities for him to pass on her post, delivered to him at the family home, and to supply her regularly with skin creams as he is a dermatologist.

The screenshots below (Fig. 4.3) show snapshots of the siblings' messages, which often revolve around the organisation of these hospital trips. Maiko-san will typically remind her brother of the next appointment and will follow such messages with a sticker depicting a bowing sheep, used to express her gratitude for his help. In the screenshot on the right, Maiko-san uses the phrase '*itsumo suimasen*', meaning 'thank you for your continued support'.[49] Interestingly she does not use the bowing sheep sticker when writing this phrase, so we can infer that in the other screenshots the sticker is doing the work of expressing this sentiment of humble thanks visually.

Maiko-san feels that stickers are a convenient shorthand for efficiently communicating what would be cumbersome to type out in words. She often uses the dictate feature on her phone to write messages rather than bothering with the smartphone keyboard, which she finds too small and cumbersome for regular use. For Maiko-san, stickers are simply another way to express oneself in a more efficient way. She feels that this is why her brother also replies with often just one sticker. She knows he is a busy doctor and therefore his replies with stickers that communicate 'roger that', or 'understood' accord with his busy lifestyle. In this way, messaging through LINE allows for a more efficient organisation of care. It also allows Maiko-san to continue to rely on her brother while communicating her sincere thanks and minimising the sense that she is becoming a burden to him.

Figure 4.3 LINE conversations between Maiko-san and her brother. Screenshots shared with permission of anonymous research participant.

One of the reasons why women may be refusing traditional family models of care based on co-residence, and may be more open to technological forms of intimacy, could be because of historical conditions that meant the home became a site of scrutiny. In developing relationships outside of this physically proximate social setting, they are seeking a kind of freedom. Hayashi-san, a woman in her mid-fifties living in Kyoto, explained that when her mother married and moved into her husband's family home in the south of Kyoto she was bullied. Her mother suffered both from the immediate family, who lived with her in the house, and by the extended family, who lived in a close-knit community in surrounding houses. She had felt suffocated by the constant monitoring of this conservative community. 'We [women] can't do anything bad because there's someone always watching what you do,' explained Hayashi-san. Her mother had passed away ten years or so ago and now she lives alone in the family home, having never married or had children. Hayashi-san channels all her difficult feelings about home into her quilting pieces, which depict *machiya*-style traditional homes. She teaches other women to do the same as a way of releasing negative associations they might have about their family homes, as described in the film below (Fig. 4.4).

Hayashi-san found a supportive community from all around the world for her quilting work on Instagram, and feels more accepted there than by conservative art societies in Japan. However, rather than seeing her online community as fulfilling all her needs, she believes that as one gets older it is important to have a supportive community living nearby,

Figure 4.4 A short video about Hayashi-san's quilt-making practice. https://youtu.be/LjReNchkCkQ.

stating that 'the internet is not your neighbourhood'. Hayashi-san volunteers at the local school and is eager to meet people at temple events. She avoided the kind of domestic scrutiny to which her mother was subject by never marrying, but she feels that ageing without any local social connections is not ideal, especially as a single person. Her digital presence, in combination with her community activities and her teaching of quilting, provide a sense of connection and mutual support that Hayashi-san feels will sustain her into the future.

Anthropological work on Japanese neighbourhoods tends to take a sceptical approach towards naturally persisting community sentiment, while simultaneously recognising locality-based 'community' as 'compelling fiction if not social fact'.[50] Anthropologist Eyal Ben-Ari was surprised to find that in two distinctively suburban sites near to Kyoto, where identity attached to locality was limited, community emerged through participation in voluntary groups with strategic aims, which he defines as communities 'of limited liability'.[51] Similarly Kavedžija found that people in a downtown Osaka community were hesitant to return to the overclose forms of neighbourhood community of the past. Instead they were keen to maintain connections to local people through participation in a community salon in which they could come and go as they chose, interacting with others in a way that balanced politeness and care.[52]

Many people shared a sense of Kyoto being particularly distinctive for the intensely observant yet highly distanced form of sociality among neighbours. It was common to hear people talk of how they felt subject to scrutiny by others, and also of their awareness of perpetuating this behaviour. Kyoto-born Akiko-san, a single woman in her sixties who lives alone, explained:

Akiko-san: You know, Kyoto is really harsh. People are always watching and criticising you.

Laura: Do you feel that women are more harshly judged here?

Akiko-san: Of course, women have it harder than men.

Laura: Who are they being judged by?

Akiko-san: Other women, older women (*obaachan*)! Because I have that in me too! To judge. It grows in me. Because that's Japanese culture. That makes Japanese people more disciplined, so it's both good and bad. I think it's very complex isn't it?

Laura: As you have become older, do you feel more free?

Akiko-san: No, the eyes that surround you will be around you for ever.

This kind of public scrutiny is matched by a certain detachment whereby people do not show their true feelings. The couple in their late fifties from the dinner group mentioned above, who had migrated to Kyoto from a rural area, explained how they had struggled to make friendships in their adopted city. They explained their difficulties by drawing on how the history of Kyoto might have led to residents being defensive against outsiders:

> They want to distinguish their enemy from the outside. I think that's why they want to protect [themselves]. At first they [Kyoto people] are very polite and kind. But after you have a relationship for a long time, their attitude changes … because they don't want to show a deep side to them. If you go into their deep side, they get angry. Their face doesn't get angry, but they are inside … that's Kyoto style.

This is a particularly Kyoto-style form of public facade (*tatemae*), long considered by scholars of Japan to be central to Japanese sociality more generally,[53] which I discuss in Chapter 5. This disconnection through public facade affords a certain freedom from scrutiny. Such a defence is necessary – not only in Kyoto but also in rural areas, where sociality can also be perceived as 'too close', especially for migrants.

Naomi-san is in her late forties. She moved with her family to rural Tosa-chō to escape a sense of being overwhelmed by 'too much information' in her previous urban hometown. By 'information', she was referring not to digital overload, but to the way in which she felt that she had to do things a certain way just because other people were doing so. In the countryside she felt that she could pursue a more autonomous life for herself and her family, enabling them to be more self-sufficient. Such a sentiment was common among migrants and is explored further in Chapter 7. However, living in a rural area also brought its own social pressures. The longer one lived there, the more the autonomy that Naomi-san desired was threatened by over-closeness:

> You know everybody here! Anywhere you go, supermarkets, petrol stations, EVERYWHERE … I can't cope with it sometimes. It's so strange not to have strangers. Here some people have never been anonymous before because they know everyone … some people even know your car number plate too!

This kind of over-closeness based on proximity led Naomi-san to adopt a specific strategy with regards to her social media presence, one that I call 'tactical invisibility'. Here she explains her approach:

> I don't really post anything and I don't look at other people's posts too because I know what they're doing! Also, what if I go to an event and there's another one at the same time? If I post about the event I went to, people will talk behind my back and everything is just so troublesome (*mendokusai*). Especially if I'm supposed to be somewhere but I don't go to that and go somewhere else … I find myself telling my friends, 'don't take a picture of me being here or post anything about it!' So maybe I'm not suited to live here so much [laughs].

This type of calibration of social proximity through social media has been theorised by digital anthropologists as 'scalable sociality',[54] in which people's media choices now afford them the ability to scale the size of audience and level of privacy according to their communication preferences. Based on my evidence in Japan, in which it was not just small group sizes and privacy that people desired, but also anonymity and invisibility, I suggest 'tactical invisibility' via the smartphone as one way in which people manage the intensities that can come with social life in small rural communities and urban centres in Japan. Visibility with regard to social media has been theorised as a way to gain recognition in the public space,[55] with a lack of visibility akin to obscurity and 'a kind of death by neglect'[56] and algorithms posing the threat of invisibility.[57] However, I argue that in becoming deliberately invisible online, people are in a better position to deal with intense forms of sociality offline.

Naomi-san is among the youngest of my interlocutors but, as we saw with Akiko-san, the pressure of visibility ('the eyes that surround you') does not diminish with age. Visibility can be mediated and somewhat controlled through digital means, however, just as scholarship has shown with regard to refugees' media practices to make themselves invisible to the state.[58] My evidence suggests that because women in Japan are subject to particularly intense forms of surveillance, digital practices of invisibility are emerging as central to female sociality. This is a direct consequence of the specific conditions within which they experience themselves within society.

Conclusion

The chapter began by describing forms of mutual support (*yui*) that were historically required for agricultural practices in rural Japan.[59] The ethnographic vignettes demonstrate the continuing importance of *yui* in contemporary Japan, suggesting that it represents a propensity towards sharing and co-operation in the face of perceived weakening 'horizontal' relationships in Japan's 'relationless society' (*muen shakai*).[60] The continuity of *yui* as socially meaningful outside of agricultural practices can be explained in relation to sociologist Pierre Bourdieu's notion of 'habitus', in which socialised norms become enduring patterns of behaviour.[61]

I have shown how *yui* is exhibited in various forms of volunteering and civic action in both rural and urban settings; it may appear both online and offline and is often expressed in gendered ways. I have attempted to show how reciprocity is at the heart of practices of dwelling, creating communities in which people (and indeed religious figures such as *Ojizō-sama*) can be cared for. By using the term 'dwelling' I am invoking anthropologist Tim Ingold's work, which highlights the ways in which people and their environments are intertwined through material and bodily practices.[62] This recognition of the ways we make meaningful connection based on proximity has important implications when considering the rise of the digital spaces in which we now live.

As is discussed further in Chapter 5, I argue that *yui* is also at the heart of how we live digitally. The rise of visual digital communication creates an affective atmosphere in which horizontal relationships and reciprocal care can be performed. The smartphone thus becomes a space of dwelling precisely when it facilitates this mutual exchange, something so central to experiences of a 'good' old age. Selecting a particular LINE sticker to express humble thanks towards a family member communicates a feeling of gratitude, yet it is also efficient and allows for a similarly brief visual response. Digital platforms become tools in the reduction of one's sense of placing burden on others or causing trouble (*meiwaku*) – both important in the context of a society where people are highly concerned by their impact on others.

I have attempted to demonstrate how the materiality of housing and neighbourhoods is implicated in the way that people think about their changing communities. Apartments designed for single people are seen to be weakening community bonds. In a parallel process I suggest that an uneasiness with the 'convenience' of smartphones is also grounded in a nostalgia for close physical proximity with which the smartphone is seen to be interfering. However, when people do experience such proximity, it can be overbearing and unwanted, as seen in the examples of Akiko-san and Hayashi-san in Kyoto, as well as that of Naomi-san, the migrant in Tosa-chō.

I found this sensitivity towards over-closeness to be an emotion particularly associated with women. In Chapter 5 I show how digital visual affordances of smartphones are enabling people, especially women, to calibrate their closeness with others more successfully, including through the practice of digital 'emotion work'.[63] I suggest that horizontal relationships, the foundation of *yui* as practised in the form of community groups, represent a kind of middle ground. In such a setting people are able to maintain connection with others and to feel that they are engaged with their communities, yet also to allow for social distance.

Yui can be understood as being dependent on physical proximity, as evidenced by community groups that seek to re-establish a sense of mutual support through regular face-to-face neighbourly contact. However, in practice, mutual care at a distance is being practised daily through the smartphone. This development is discussed further in Chapter 5 in relation to friendships and in Chapter 6 in relation to preventative health measures. Rather than being rooted in physical proximity, the smartphone is emerging as central to a kind of care that provides the benefits of community, including a continued sense of *yui*, without some of its more oppressive associations.

It is also important to recognise the negative implications of the rise of digital platforms for community care. For example, for individuals who are unable or unwilling to adopt technologies such as the smartphone, there is a risk that this shift may exacerbate feelings of isolation and loneliness. Furthermore, the dependence on smartphones for community may also amplify concerns related to privacy, data security and the commodification of personal relationships in the digital age.

I found that digital practices are in effect being used to mediate a balance between dependency and the desire for autonomy, as argued to be central to social relations among older adults in Japan more generally by anthropologist Iza Kavedžija.[64] The smartphone reduces the need for physical proximity and gives people more choice as to when and where they connect. To a degree, this affordance of smartphones sits alongside the work done by *tatemae* or one's public facade, which allows people in close proximity also to maintain a distance within face-to-face interaction and to deal more successfully with the sense of scrutiny that people felt, even when living in urban contexts such as central Kyoto.[65] In the next chapter I develop the idea of 'tactical invisibility' in relation to anonymous uses of social media that I observed to be in operation, in order to insert distance into familial relationships.

Notes

1. Nakane 1967, 1970; Ariga 1957.
2. Suehara 2006.
3. Sekine 2023.
4. Love 2013.
5. The Government of Japan 2005.
6. Norbeck 1977.
7. Ben Ari 2013.
8. Schwartz 2003, 45.
9. Atsumi and Goltz 2014.
10. Aldrich 2011.
11. Tatsuki 2000.
12. Daniels 2022.
13. Hjorth and Kim 2011.
14. Nakano 2004.
15. Taniguchi 2010.
16. Okabe et al. 2019.
17. Nakano 2004.
18. Firth 2020.
19. Onitsuka and Hoshino 2018.
20. Onitsuka and Hoshino 2018.
21. Traphagan 2004.
22. Rosenberger 2006.
23. Nakane 1970.
24. Hata and Smith 1986.
25. Befu 1993; Goodman 2005.
26. Suehara 2006.
27. Goodman 2005.
28. See Clammer 2010 for a discussion of *kanjinshugi*, referencing Japanese scholars such as Hamaguchi 1982; Murakami et al. 1979.
29. To hear a locally renowned singer perform a traditional rice harvesting song, see the video in Chapter 1, 'Introduction to Kōchi'.
30. Allison 2017.
31. Ozawa-de Silva 2018.
32. Allison 2013.
33. Allison 2013, 15.
34. Horiguchi 2011.
35. Eiraku 2019.
36. Tateno et al. 2019.
37. Kavedžija 2019; Kavedžija 2018.
38. Kavedžija 2018.
39. Pekkanen et al. 2014.
40. Ben-Ari 2013.
41. Hendry 1993.
42. Hjorth and Kim 2011, 188 (original emphasis).
43. Nozawa 2015.
44. Nozawa 2015, 377.
45. Nozawa 2015, 380.
46. Miller 2008.

47. See Traphagan 1998 for a discussion of *meiwaku* among older Japanese adults.
48. A comic developed with Maiko-san is presented at the beginning of Chapter 9, along with a further discussion of her technology use.
49. '*Sumimasen*' is generally used to express thanks to someone who is actively going out of their way to help. It is more humble and polite than a simple thank you (*arigatō gozaimasu*) and acknowledges the work of someone else. Here Maiko-san uses the slightly less formal '*suimasen*', but this still has the same connotation of being very grateful.
50. Kelly 1993b, 401.
51. Ben-Ari 2013, 11.
52. Kavedžija 2018.
53. See for example Hendry 1989; Sugimoto 2020; Naito and Gielen 1992; De Mente 2005.
54. Miller et al. 2016.
55. Thompson 2005.
56. Thompson 2005, 49.
57. Bucher 2012.
58. Ullrich 2017.
59. The degree to which such co-operation around rice farming should be taken as foundational to wider political systems in Asia was a classic debate in the history of anthropology. Historian Karl Wittfogel in the 1950s famously argued that the necessity for co-operation in rice farming was the very reason why China and other 'oriental' societies developed despotic states (Wittfogel 1957) – an argument subsequently repudiated by anthropologist Edmund Leach's ethnography of rice farming within a Sri Lankan village (Leach 1961).
60. Allison 2017.
61. Bourdieu 1977.
62. Ingold 2000.
63. Hochschild 1979.
64. Kavedžija 2015a.
65. The use of digital communication to develop a sociality that is regarded as neither too cold nor too warm is not restricted to Japan, but has also been observed in a suburban English context by anthropologist Daniel Miller (Miller 2016).

Image by Laura Haapio-Kirk.

5
Crafting the smartphone: visual digital communication

> I think when we are elderly, we may not have friends right next to us. So the smartphone might feel more precious to us because it allows us to stay sociable.
>
> Kazuko-san, aged 58, Tokyo.

Kazuko-san lives in Kyoto and has friends across Japan. She feels that the smartphone will become increasingly important to her as she ages since it will keep her friends close. Currently the smartphone helps her to stay in touch with her friends, even though she does not live near to them. Having her friends always available to message through the application LINE gives her a sense of comfort and strength when dealing with difficult things in her life, such as caring for her elderly parents and mother-in-law, or when she herself became ill. 'I felt that they (my friends) were next to me all the time.' For Kazuko-san, the smartphone became a device through which a feeling of proximity with her friends could be achieved, and through which care could transcend distance.

While people in Japan have had longer to get used to the concept of internet-enabled mobiles than elsewhere, as described in Chapter 1, the rise of the smartphone for many people has introduced a new speed and casualness to textual conversation, making it feel more like a face-to-face conversation. The visual communicative affordances of the platform LINE, with suggested emojis appearing as you type and with stickers that efficiently convey meaning without typing long messages, mean that conversations can flow very fast, even among recent smartphone adopters. Before switching to smartphones, groups of friends would stay connected via their feature phones (*garakei*) through long email chains which were hard to keep track of – and, more importantly, did not convey the warmth of their friendship. As Kazuko-san explained:

> Before, communicating with each other using the *garakei* was more business-like, it wasn't full of personality and liveliness like with a

smartphone. Now you don't need to be formal and write an email subject, you can just check in on your friends with a quick message or sticker. It's more of a casual conversation.

'Checking in' and sending quick messages offers a baseline of sociality which, while recognised by participants as not being the same as face-to-face interaction, still conveys care through attention; it provides the kind of casual sociality that would typically come with close physical proximity. Anthropological definitions of care are diverse, but the theorisation of care as conveying and producing 'the way someone comes to matter', by the anthropologist Lisa Stevenson,[1] is helpful here in dealing with the digital forms of attention and affect that I observed. Visual digital communication can also be a powerful tool for the performance of 'emotion work',[2] something that is also important in the kind of peer-support sought and found in female friendship groups such as Kazuko-san's.

This chapter examines how 'digital repertoires'[3] among older adults in Japan are developed in correspondence with practices of care. Within a Japanese social context in which much attention is paid to 'correct' forms of sociality, whereby the mode of communication holds as much significance as the content,[4] I argue that a proliferation of visual digital communication has emerged as an increasingly important part of the digital repertoires of older, particularly female, smartphone users as a mediator of care.

The anthropologist Marjorie Goodwin argues that it is in embodied acts that communicate affect, attention, empathy and co-presence that care is realised.[5] I extend these ideas of embodiment and proximity as being central to care by exploring how visual digital communication can create an aesthetics of care that draws on embodied expression. I pay particular attention to the use of LINE stickers as a mechanism for balancing both closeness and distance that emerges as necessary in relationships of care, in line with anthropologist Iza Kavedžija's observations of relationships between older adults in Japan.[6] The recent 'affective turn' in the humanities and social sciences has fostered a growing recognition of the importance of paying attention to 'affect', i.e. recognising how emotion, feelings and affective states are intertwined with social and cultural life.[7] In this chapter I employ the notions of 'affective labour'[8] and 'emotion work'[9] to focus on visual digital practices, arguing that emojis and stickers have become central to the performance of 'care at a distance'[10] in the way that they facilitate digital 'atmosphere'.

By adopting a 'digital repertoire' focus, I follow anthropologists and technology researchers Hänninen et al. in approaching the use of digital technologies in a holistic way, recognising how personal interests, skill sets and availability of technologies combine to shape the way in which older adults develop repertoires of use to meet their daily needs.[11] Digital repertoires, like media repertoires, can be understood as 'a set of meaningful practices'[12] which are developed in social contexts.[13] I draw on this understanding in my analysis of the ethnographic examples below to show what older adults do with the technologies they use, and how they creatively deal with problems such as the effects of lack of physical proximity on communication. Rather than focus on digital divides or how older adults become digitally skilled as they adopt the smartphone, this chapter demonstrates how emerging digital repertoires can allow older adults to mitigate their sense of vulnerability and dependency on others as ageing persons, while simultaneously embedding them in relationships of care. I also present the more challenging side to smartphones, including a fear of being misunderstood, which I show is also somewhat mitigated through visual digital practices.

The mobile landscape

Discourses around the development of mobile technologies in Japan have followed the wider trend towards 'techno-nationalism', as also seen with robotics, situating technological advancement within the rhetoric of national reconstruction.[14] However, Japan has been slow in many ways to adopt new technologies, despite an international perception of being at the forefront of technological innovation. Daily life in Japan is remarkably low-tech: the economy is still largely cash-based despite government attempts to encourage card and mobile payments. Much bureaucratic work is still paper-based rather than computerised and old technologies such as fax machines remain popular. Indeed, the name of the new *Reiwa* era was announced internationally in April 2019 via a fax sent to Japanese embassies worldwide.

Japan had initially led innovation in mobile communication. Mobile phones in Japan have been distinguished by their provision of internet access – in other words, they have been 'smart' – for a lot longer than elsewhere; internet-connected feature phones, known as *garakei* (Fig. 5.1), became popular a decade before the advent of the smartphone.[15] Mobile phones first appeared in Japan in the late 1980s, operated by Nippon Telegraph and Telephone (NTT), a public corporation controlled by the government.[16] The devices were largely in demand only by corporations, given their high cost, and it was not until the late 1990s that the private

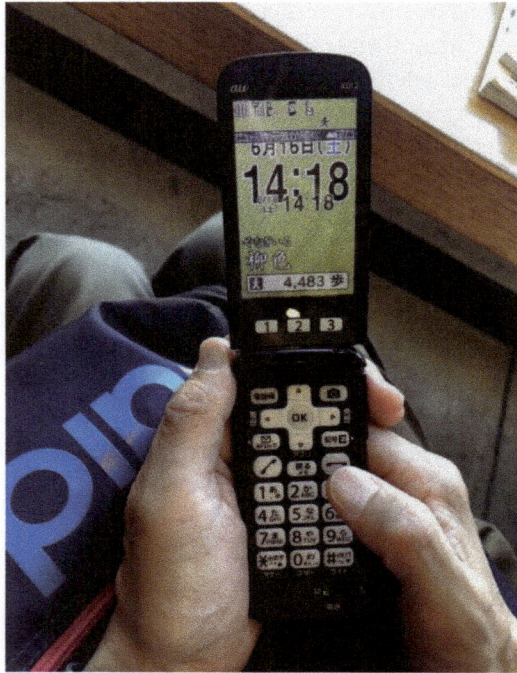

Figure 5.1 A research participant holding his 'feature phone' (*garakei*). Image by Laura Haapio-Kirk.

spin-off company NTT Docomo launched packages intended for individuals.[17] The demand for mobile phones was particularly strong among young people, building on the popularity of pagers for casual communication among the youth at the time.[18] '*Keitai*' (mobile) culture was thus shaped by the creative usage of young people, marking the birth of the mobile society in Japan.[19] However, my research indicates that smartphones are now reaching all demographics of society – not through top-down efforts, but typically because middle-aged people are equipping older relatives with smartphones so that they can care for them better at a distance.

The world's first commercial mobile internet service appeared in Japan with the introduction of iMode, launched in 1999 by NTT Docomo.[20] This service was quickly adopted, reaching 70 per cent of all mobile users by the mid-2000s,[21] likely because of the specific context in Japan at that time of low rates of PC usage.[22] Internet-enabled feature phones are popularly termed *garakei*, a mix of the word for mobile phone, *keitai,* and the shortened word for Galapagos, *gara*. The name took inspiration from the 'Galapagos syndrome' in which isolated species evolve distinctive features, referencing the perceived uniqueness of the Japanese mobile landscape.[23]

Garakei typically have voice calling, emailing and text messaging capabilities, in addition to basic multimedia and internet access. These devices are often flip-phone in design, featuring a physical keyboard. My older research participants particularly valued their hard-wearing design, long battery life and non-touch screen interface. They also liked the simplicity of the *garakei*, with voice calling, messaging and step counter being among the top three uses for their phones. Messaging (*meeru*) via a *garakei* is typically conducted via email instead of SMS messages, and every phone is sold with an associated email address that a person creates when establishing a contract.

While the popularity of *garakei* was initially driven by adoption from teenagers, who shifted the association of the mobile phone with business to teen street culture,[24] my research in the smartphone era has found that *garakei* have become synonymous with older mobile phone users. However, during the period of my ethnography between 2018 and 2019 on which this book is based, many older people switched to using a smartphone – making this a perfect time to assess the impact of the smartphone on their lives and to understand why some were hesitant to give up their *garakei*. Many of my participants had both a *garakei* and a smartphone; others had only *garakei* and a few people only had landlines.

Smartphone adoption in Japan accelerated after the 'triple disaster' of 11 March 2011 (known as 3.11), a consequence of the immediate rise of information sharing on social media such as Twitter.[25] The short video below (Fig. 5.2) shows how the notion of the smartphone as a lifeline remains strong in disaster-prone Japan, particularly among older adults living alone. Smartphone ownership among seniors has rapidly increased over the past few years as the government put pressure on telecommunication companies to reduce the price of packages to bring them in line with other countries. Shortly after I had left Japan in the summer of 2019, friends reported long queues outside mobile phone shops as alluring promotions were applied to smartphones.

Several of my participants switched to smartphones at this time. Among them was Maiko-san, a woman in her sixties from Kyoto who I introduced in Chapter 4. She had been previously adamant that she would never get one, but she was persuaded to after a large typhoon struck Kyoto. 'Guess what? I finally got a smartphone. I like it!' she immediately messaged me via Facebook Messenger after acquiring her new device. She had previously resisted as she did not want the additional

Figure 5.2 A short video about how the smartphone can be a lifeline. https://youtu.be/z9AY2KWj4Rs.

social pressure of being 'always available' that she felt acquiring a smartphone would bring. In addition, she felt that her poor eyesight would hinder her ability to write messages with such a small keyboard, preferring to use social media on her home computer. So when she discovered the voice-to-text feature on her new phone she was particularly happy. As I was now back in the UK after returning from fieldwork, we collaborated on the video above as a way to engage in remote ethnography.[26] The video emphasises how the smartphone brought her a sense of reassurance as an older single woman living alone, particularly in times of crisis. After the typhoon she had no electricity at home and therefore no way of accessing news via her TV or computer. 'I felt like I was blind,' she said. This sense of vulnerability is what prompted her to switch. The Covid-19 pandemic has further accelerated uptake of smartphones among older demographics in Japan,[27] demonstrating their value as a safety net in uncertain times.

During my fieldwork it was common for people to own both devices, using their *garakei* for phone calls and their smartphone for the internet.[28] However, this transition period may prove to be protracted. While I was in Japan one woman in her sixties received a letter from her phone company telling her that her *garakei* would soon be unable to receive updates and that she must switch to a smartphone, thus demonstrating the power of mobile phone providers to influence consumer behaviour. A typical monthly smartphone bill among my research participants during my fieldwork was between ¥7,000 and ¥10,000 (£53 and £75), which in rural Kōchi could be a month's rent for an apartment.

My ethnography situated the financial burden of smartphone adoption within a range of other factors that influence whether someone switches to a smartphone. These may include an individual's openness towards trying new technology, their attachment to a network of contacts perceived to be contained in the *garakei* and sometimes a simple matter of not wanting to let go of a seldom-used phone number – possibly without the realisation that you can transfer numbers to new providers. Sometimes, if people did not want to change their mobile number, they bought an iPad to use LINE (by far the most popular social media platform in Japan), often through the same phone company that would supply their WiFi. In one instance a woman in her sixties kept her *garakei* purely out of feeling bound to her phone company because of the 'family plan' she was on; this meant that she could send free messages to her family. However, she recognised the futility of maintaining her *garakei* contract since people were now mostly contacting her for free via social media on her smartphone. There was a kind of liminality between the *garakei* and smartphone that was common among my research participants, as expressed by this woman:

> It's not necessarily cheaper to call on the *garakei*. I can't leave my *garakei* because I don't want to leave Docomo [phone company]. But these days everyone calls me on Facetime or LINE. You don't need a phone number any more. Sometimes, only sometimes, people call me on the phone number.

Visual digital communication

Interpersonal communication has long been a key research area in social anthropology.[29] Recently, anthropologists have turned to the profound impact of social media on social relations.[30] In particular, digital visual communication is recognised as an effective and accessible form of communication,[31] with an increasing number of studies in the fields of digital anthropology, media studies and internet studies exploring the consequences of digital images on sociality.[32] The shift towards visual digital communication was already felt 20 years ago to be heralding a new era in terms of literacies and relations of power with regards to the democratisation of representation.[33] Paying attention to the visual nature of sociality has become imperative as social media has become a ubiquitous part of the lives of most people around the world, demanding innovation in anthropological thought and methods.

The most popular app in Japan today is a messaging service called LINE. It was initially launched in 2011 by South Korean company NHN Japan as a messaging service for their employees to communicate following the 3.11 earthquake and tsunami, which had severely affected Japan's telecommunications network. During the days following the earthquake and tsunami phone lines stayed down, but data channels remained open, becoming the most effective way to stay in touch.[34] The application was subsequently released to the public in June 2011; by 2013 it had become Japan's most popular social networking service.[35] By 2018 it had reached 78 million Japanese users[36] and 165 monthly active users worldwide,[37] serving major markets outside Japan such as Thailand, Taiwan and Indonesia.

LINE was the most commonly used app among my research participants in both Kyoto and Kōchi. I came across no one with a smartphone who did not use the application, and some even referred to their smartphone as 'my LINE' *(maiLINE)*, paralleling discourse that accompanied previous waves of personal technology adoption, including in the 'my car' *(maikuruma/maika)* and 'my colour (TV)' *(maikara)* era of the 1950s and 1960s.[38] During these early years of the consumption boom, the borrowed possessive term *'mai'* was placed in front of newly acquired items, reflecting how owning these objects secured an individual's place in consumer society. Wada-san, aged 75, explained how she came to acquire *maiLINE*, revealing how this particular consumption choice reflected her embeddedness in social relationships rather than being individualising:

> I actually bought my LINE this Sunday because my daughter told me to. I use LINE to contact her almost every day. This is so easy to get in contact with people! I use the microphone to message as well, which is a great feature.

For such people, the entire concept of a smartphone was conflated with the use of LINE. Predominantly people used the application for messaging and free voice and video calls, but LINE is a 'super app' mega-platform.[39] It has a range of features such as a mobile wallet, on-demand video, news, music and manga, as well as user profiles which can be updated with statuses, music and profile pictures. While LINE offers many functions, making it closer to the Chinese application WeChat than Facebook-owned WhatsApp in terms of its multifunctionality, I did not observe these additional features being widely used by my research participants. They would use Instagram or Facebook to share photos and post status updates, and a few had specific applications for making mobile payments, such as PayPay. A small number of individuals did use LINE's wallet feature, but only because of a promotional offer in which they could win money back every time they paid for something at the convenience store FamilyMart using LINE Pay. As one woman in her fifties observed, 'It's like a lottery but you never lose'. Among younger people there appeared to be a fuller use of the spectrum of features on offer, with younger users reading manga and adding playable songs to their profiles.

One of the most distinctive features of LINE is the highly visual nature of messaging through a plethora of illustrated emojis and stickers (*sutanpu*). It was visual messaging that emerged as distinctive of the digital repertoires of older, particularly female, participants in this research. This section will explore some of the reasons behind this observation. Japanese older adults have been adopters and innovators of computer mediated communication (or CMC for short) since the early days of the internet, as the digital ethnography of a senior virtual community undertaken by media studies researcher Tomoko Kanayama demonstrated.[40] Kanayama conducted participant observation on a mailing list between 1999 and 2000. He discovered that the 120 members of this highly active group overcame limitations of text-based communication by experimenting with a variety of language forms including haiku and emoticons. The research revealed the propensity of older Japanese adults to find ways of overcoming the lack of physical proximity that might otherwise afford emotional cues in digital communication. This early research has resonances with the way in which I found older adults used digital communication via the smartphone to practise care at a distance.

'Emoticons' are sequences of keyboard symbols expressing facial expressions, such as :-). In Japan these early digital carriers of affectivity evolved into *kaomoji* (literally meaning 'face character'). They were made up of characters and punctuation marks, used to express more complicated facial expressions or gestures such as m(_ _)m which depicts a person bowing. The 'm' on either side of the brackets indicates hands, with a head pressed to the floor in the middle. The image below (Fig. 5.3) shows an evolution from kaomoji to emoji to sticker, all showing the same sentiment. 'Emojis' refer to more pictorial and graphic symbols made with Unicode. Emoji is a compound Japanese word meaning 'picture-character', with the 'e' referring to picture and the 'moji' to character or letter.

In the 1990s, the first emojis were created by a Japanese telecommunications worker called Shigetaka.[41] Japan has since been at the forefront of visual digital communication, such as the development of emojis and stickers – inevitably drawing on the visual traditions of manga that dominate the aesthetic of much public and private discourse. From posters on

Kaomoji	Emoji	Sticker

m(_ _)m ⟶ 🙇 ⟶

Figure 5.3 Evolution of the bowing emoji. Image by Laura Haapio-Kirk.

trains to information booklets in doctors' clinics to private messaging, the conventions of manga are omnipresent in visual modes of communication in Japan. Visual digital communication can be understood as also emerging from the entrenched pictographic nature of the Japanese language. One of its three writing systems, 'kanji', is ideographic in nature, meaning that a symbol is used to represent a meaning or idea.[42]

'Stickers' are the latest iteration of visual digital communication. Typically larger than emojis, they are sent as individual messages rather than embedded within text. Stickers display much more diversity in terms of expressive ability, styles and animation. On LINE there are two main types of stickers available: custom stickers and downloadable sticker sets. Sticker sets usually contain between 16 and 24 individual stickers, tied together by a cohesive theme. LINE stickers can be personalised in various ways, including through the purchase of tailored sticker sets with a person's name embedded within the stickers. Such tailored sets usually cost the equivalent of between £2 and £3. They are sometimes given as virtual gifts between friends, though most of my participants were cautious about spending too much on stickers, no matter how 'tempting' they might be. As of April 2019, approximately 5 million LINE sticker sets were available and stickers are sent on an average of 433 million times a day.[43]

There is abundant research on the communicative affordances of emojis,[44] demonstrating the ways in which visual digital messages are used to articulate non-verbal cues,[45] thus conveying particular attitudes.[46] As mentioned above, images are central to the Japanese language in the form of *kanji* (literally meaning 'Chinese characters' in Japanese) which are ideographic and rebus characters.[47] The rise of digital visual communication, such as emojis and stickers, can therefore be seen to stand in continuity with prior forms of written communication.

Their difference lies, I argue, in that stickers often serve a phatic[48] purpose, as their social function is more important than their 'meaning'. However, some do have text embedded within them. Indeed, their meaning is often ambiguous. It is the open-endedness that the visual affords that makes them more akin to how my participants would communicate offline, their meanings often being implicit rather than overtly stated. Like digital photography, which has been shown to be important for visual phatic communication in daily life,[49] I contend that stickers predominantly serve to maintain social connections rather than being used because of their inherent meaning. As I will explore in the case below, this social function can be understood as constituting a digital 'atmosphere' so crucial to offline communication.

Most smartphone users around the world are likely to be familiar with emojis, and now even with stickers, which have infiltrated platforms such as Facebook Messenger and WhatsApp. What my research revealed is the way in which stickers afford communicative possibilities that people are particularly concerned may be missing in textual messaging. First, many people explained that the use of stickers has made messaging more 'conversational' because of the ease, speed and frequency with which stickers can be sent. While many of my research participants would spend time carefully checking their textual messages for typos and trying to avoid any possibility of miscommunication, for example, they would swiftly choose stickers, which better conveyed a wide range of meanings with ease.

Stickers excel in their capacity to convey sentiments that can be challenging to articulate, featuring characters showcasing intense emotions spanning from profound sorrow to subtle hostility to ecstatic joy. The Japanese language is multi-layered and often indirect, so meaning is hard to convey through text alone, without the accompanying tone of voice and body language. Hall proposed that Japanese is a 'high context' language, in which significance is implicit; a listener is expected to be able to read 'between the lines' to comprehend meaning which is contained in the physical context or internalised in the person.[50] As one man in his forties in Kōchi explained, visual communication via LINE stickers offers one way around this problem:

> People here [in Japan], they have subtle feelings and sometimes people say, 'you are stupid!'. But there is 'YOU are stupid!' or 'you ARE stupid', right? Different tones. So stickers can show that kind of thing much better than text. The best thing is seeing each other, face to face.

Later in this chapter I will return to a discussion of the ways in which visual digital communication can mitigate some of the anxieties caused by the lack of physical proximity in digital communication. For now I will demonstrate through case studies the ways that people use stickers to craft their social interaction in various contexts.

Visual communication and age

At one lunchtime meet-up, Kazuko-san and her friends explained that while they are all middle-aged, most of their co-workers at the catering company they work part-time for are younger. At work the younger generation are bound by cultural and linguistic rules, requiring them to speak to older colleagues using language that conveys respect, formality and a degree of distance. However, on the work LINE group these boundaries are transgressed. In LINE messages, primarily using stickers, meaning can be conveyed with minimal words, so normal linguistic rules for cross-generational communication can be relaxed. Stickers which have text embedded within them were especially effective, removing the responsibility of having to decide on an appropriate level of formality in language. Kazuko-san explained:

> I think that is when stickers are quite useful, when speaking to different generations. Because stickers make communication more approachable, warmer. You can use them across all ages. It's not that this makes us feel younger but more that it makes us not even think about our age.

By not having to conform to traditional modes of formal language, the selection of a particular sticker to express a phrase or a sentiment is a kind of leveller. It serves to bypass modes of conduct otherwise required and to 'warm up' otherwise cool relationships, meaning that people are not even considering their place within the usual age hierarchy which otherwise structures their interactions. Stickers often have language embedded in them that is more relaxed and common between friends, rather than the formal discourse used between older and younger (*senpai/kohai*) members of a workplace.

One of the women frequently received stickers from her mother, for example to say 'good morning' and to keep in touch throughout the day. Her mother had independently downloaded a sticker set featuring a humorous grandmother character with whom she felt she identified. The embedded text in the stickers was casual in form, for example including the text 'ryo', a slang version of 'ryokai' that means 'confirmed'. In this choice one can see that stickers can be used to play with age-based categorisation, embracing caricature at the same time as levelling age-based differences in styles of communication. Similarly Keiko-san, who was introduced in Chapter 1 and lives in Osaka, downloaded a humorous sticker set featuring a middle-aged woman, along with embedded text written in the local Osaka dialect (*Osaka-ben*). Using this sticker (Fig. 5.4), which she only used with certain female friends, Keiko-san could reveal a more playful side to her personality and use self-deprecating humour that played on a stereotype about Osaka older women.

Figure 5.4 LINE conversation featuring a sticker of a stereotypical middle-aged Osaka woman character. Screenshot by anonymous research participant.

Discourses of ageing around the world tend towards collective identities for older adults that imply increasing dependency and otherness with age;[51] they frame older people as passive recipients of care[52] and situate them outside mainstream society.[53] Visual discourses of ageing tend to offer similarly ageist tropes, in what Loos and Ivan term 'visual ageism'.[54] Following the assertion made by the gerontologists Minichiello et al. that media representations provide a lens with which to study how ageing is socially constructed,[55] I propose that through digital visual media selected by older adults themselves we can glimpse their resistance or playfulness in response to such age-based othering.

Through the stickers in these three instances above (a humorous and feisty older woman character, a bold middle-aged character and stickers that made people not even think about their age), we can see how digital visual representation can overcome 'otherness'. Older adults often face forms of ageism and 'othering'; this can be heightened for women who are under pressure to behave in certain ways according to their age, at risk of becoming 'invisible'[56] and so becoming the most 'othered' of all. It thus seems unsurprising that it was women who I found to be embracing digital platforms and the affordances of visual communication in order to fashion a self-image to set against such stereotypes. It is older women who have grasped the potential for self-representation through the smartphone, claiming visibility that may not be possible in their offline lives.

Personality and authenticity

The 'authenticity' of technology-mediated humanity is a persistent concern, reflected not only in popular discourse but also in studies of digital communication. Scholars such as technology studies researcher Sherry Turkle have highlighted the ambivalence that many people feel about increasing dependency on digital devices, fearing that digitally mediated relationships may be shallower than those conducted offline.[57] However, my research accords more with anthropologist Dorinne Kondo's assertion that people have multiple selves,[58] contributing to a growing body of research arguing that offline relationships are no more authentic than online ones[59] – and indeed that online lives can sometimes be more 'real' in terms of personal aspirations and the feeling of 'being-at-home' than offline ones.[60] My ethnography revealed how the personalisation of visual digital communication allowed many participants to extend their personalities into their messaging beyond what would be possible through text alone.

Sticker sets on LINE tended to be chosen and downloaded because they expressed the personality of the user in some way. Multiple people purchased stickers with their names embedded within their design. Even without a name, they instantly knew which friend had sent a message by the character in the sticker:

> Look [points to Line sticker showing a cartoon dog], I can tell that is from Kazuko-san because Coco-chan is very 'her' (Kazuko-san rashii)! I think sometimes people who are cute (*kawaii*) have super cute stickers; one of our friends is super cute so she uses really cute stickers.

Kazuko-san's friend used the phrase 'Kazuko-san rashii', indicating that the sticker represents Kazuko-san but is not necessarily her. This idea of digital practices reflecting a sense of self was sometimes extended to their entire device. People often felt an increased sense of closeness and attachment towards their smartphone compared with their previous *garakei*. This is because the user has greater control over how they craft the phone's interior to match their personality. Kazuko-san explained:

> I feel more towards the smartphone than my old *garakei*. I want to have it near me always. I feel shy about showing people my apps. It's kind of like a bookshelf, it shows my personality. You would only put the covers of the books you want to show at the front too. The smartphone is who I am.

Sometimes for this reason participants were often shy about showing me the interior of their phones. For that reason I left my smartphone interviews until the end of my fieldwork, when I had established trust with people.

Visual digital media allows people to communicate in highly personal ways due to the range of emotions and expressions that they are easily able to display. Research participants could choose the precise LINE sticker to convey a feeling that might difficult to put into words. Sticker sets often encompass a wide variety of emotions, as described above. While participants always stressed that face-to-face communication was best for understanding what people truly meant, when communicating through messaging they felt that stickers were integral to creating the kind of multi-layered communication analogous to their engagement offline.

Stickers often display exaggerated facial expressions and bodily postures, such as characters showing gushing tears or blowing kisses.

Rather than approaching visual digital communication as a direct conduit to inner, and therefore more 'authentic', feelings, stickers and emojis reveal the intricacies of Japanese sociality in which emphasis is placed on the surface. Indeed, as anthropologist Philip Swift articulates in relation to the performativity of prayer at Shinto shrines, it is helpful to imagine Japanese sociality like its cosmology, as 'constituted in facades and fabricating practices'.[61] However, the performative affordances of visual digital media such as emojis and stickers have now spread beyond Japanese sociality, travelling around the world with the arrival of the smartphone in the late 2000s. What emerged in my research as distinctive of Japanese sociality is the frequency with which visual digital communication was referred to as fixing the concerns that older adults had about the inadequacies of textual messaging for managing their public facade.

The concepts of *honne* (true voice) and *tatemae* (public facade) have a particular history in the post-war discourse on 'Japaneseness'.[62] As Japanese studies researchers Goodman and Refsing note:

> one of the first distinctions any anthropologist embarking on research in Japan learns to make is between *tatemae* and *honne*. *Tatemae* ... refers to an individual's explicitly stated principle, objective or promise; *honne* refers to what that individual is really going to do, or wants to do.[63]

I recognise that these terms belong to essentialising discourse that Japanese people often have about themselves, associated with the wider *nihonjinron* tradition of Japanese identity-making that emerged in the post-war period.[64] However, these words still had meaning for my interlocutors, as became evident from the way that they were reflected in their use of smartphones. Listen to a local woodworker from Tosa-chō talk about his thoughts on *honne* and *tatemae* in the video about purpose in life in Chapter 8 (Fig. 8.1).

In a similar vein, anthropologist Joy Hendry's work in Japan reveals how 'wrapping' is important both as a material and a social practice.[65] By this she is referring to how, just as presents are wrapped and bodies are clothed, communication is enveloped in layers of polite language and non-verbal symbolic behaviour. Hendry argues that rather than being seen as mere decoration, we should take wrapping seriously as a way that people impress and manipulate one another – not just in the Japanese case, but elsewhere too. As demonstrated in sociologist Erving Goffman's[66] work on the presentation of the self, the social expectation for individuals to engage in highly performative communication that complies with collective interests or the internalised normative order of organising social life has found wide traction outside of Japan. Being aware of social etiquette, and being able to perform an expected facial expression in various situations, are regarded as essential parts of socialisation.

Visual communication facilitated by the smartphone does not reduce the 'authenticity' of interpersonal communication. Rather, it reduces the effort of living up to social expectations, which, more often than not, go against individuals' 'authentic feelings'. Sometimes these authentic feelings are difficult to locate, an experience that spans generations. In discussing a drawing that Akiko-san, aged 35, made to illustrate the concepts of *honne* and *tatemae* (Fig. 5.5), she explained, 'honne is like a shadow. You can see the shadow, but you can't touch the shadow ... you can feel the *honne*, but you can't actually know it.' In selecting the appropriate image that captures the emotion one wishes to project, people are participating in the same social performance that happens offline, but without as much required effort. Or, to use Hochschild's term, less 'emotion work' is required.[67]

Being able accurately to 'read the air' (*kūki o yomu*) by paying attention to bodily expression, and to mask problematic feelings, are an important part of Japanese social relations; they are indeed considered a sign of maturity.[68] My research revealed that a core concern behind hesitancy over digital communication among older adults was anxiety about 'reading the air' online, a skill honed in offline contexts over a lifetime. These older adults perceived textual online communication to be limited because of the impossibility of non-verbal cues that they relied

Figure 5.5 *Honne* as a shadow, hiding behind a smiling face. Drawing by anonymous research participant.

on in offline interpersonal communication. I argue that it is this concern for exchanging messages that will not be misunderstood which has given rise to the explosion in visual digital communication that we see today among Japanese older adults, especially women. I propose that LINE stickers create 'atmosphere', which is essential for Japanese social relations more broadly.

A common refrain from people of all ages, but especially older adults, was that messaging using text alone made them anxious; they felt that it was easy unintentionally to cause awkwardness or tension. This feeling was summed up by one woman in her sixties:

> If communication is not face to face, you can't express your emotions or feelings. Face to face allows more feeling ... that's the most difficult thing about the smartphone. Whenever I am typing out a message, I feel it's so hard to read the other person's emotions. Your message might sound angry, even if that's not what you mean. So I really try to be careful whenever I am sending a text. It's hard to express yourself.

The frequency with which people spoke about the limitations of textual messaging was matched by how they spoke of digital visual media, including stickers, as an aid for achieving more fluid communication online. Stickers were used to help others read the digital 'air' and to appreciate intentions and multi-layered meaning that could otherwise be lost in text-only messaging. As Kimori-san, a man in his forties, explained, it is easy to overthink what people mean when receiving short textual messages such as ones that just say something like 'OK'. Kimori-san prefers to spend time writing long messages; he also uses emojis and stickers to express feelings more clearly. As he explains, 'I'm always thinking of the worst-case scenario, so better to be more aware than to make mistakes'.

The root of the fear for many people was the ostracism that might occur if they expressed something in a way that demonstrated an inability to 'read the air'. As Yuriko-san, a woman in her sixties, explained:

> We are all living together (*minna de isshyoni*), it's a collective society (*shūgō shakai*). And the scariest thing for that kind of society is to be un-included, to be rejected and alone... Human beings live in that complex society, and so the fact that one can become isolated from society because of information (*jōhō*) is scary.

An association emerged between being able to 'read the air' and being a 'good person' who is accepted in society. While LINE helped Yuriko-san to communicate with people daily, she felt that the kind of communication she could have offline could never be achieved online. She distinguished between 'reading the air' (*kuuki*), which literally means the air in a physical sense, and atmosphere (*funniki*), which refers to the mood in a more abstract way. However, both were inherently seen as existing offline in the space between people:

> And all of that communication was made possible because of this atmosphere (*funniki*) we have between us … not only through words, but also through feeling, the atmosphere and not air (*kuuki*), but this kind of mood, like a sponge … So direct contact with people not only teaches you the skill of communication, but also makes you a better person, I think, because you are able to read the atmosphere. And that definitely can't be done in the virtual world.

Likewise, one must carefully judge the appropriateness of forms of communication according to the situation. Sometimes people felt it would be inappropriate to send a sticker as it might make the conversation too casual or even be disrespectful, depending on one's relationship to the recipient. Confucian ethics in Japan prioritise social hierarchy and social harmony in interpersonal communication.[69] Depending on the person one was messaging, people carefully crafted their messages to suit the social situation, 'polishing' their messages until they took the correct form.

Iza Kavedžija's ethnography of a community salon in Osaka revealed how such careful attention to form and politeness in face-to-face communication ensured that people felt comfortable in one another's presence as they negotiated proximity and distance in this densely populated urban setting.[70] Following Kavedžija's assertion regarding the importance of form and its relevance for negotiating proximity, in the next part of this chapter I show how people negotiate their sense of proximity through visual digital messaging. Despite its limitations when compared with face-to-face communication, my evidence suggests that the rise of digital visual communication has enabled people to express themselves in a way that invokes the kind of multi-layered sociality so integral to offline communication, based as it is on proximity and physicality.

Digital proximity

'It's really hard and sad to see your own mother and father deteriorate, especially if they get dementia. It's like a tunnel without an ending,' observed Noriko-san, a 62-year-old woman who returned to work at a catering company after her children left home. Now she is balancing the demands of work with caring for her mother, who lives with her in Kyoto. She knows that her mother would not approve of moving to a care home, and she wants to continue looking after her for as long as possible. Noriko-san regards the type of care required for parents with dementia (*ninchishō*) as rather different from caring for children, which is more emotionally rewarding. She felt the care of older adults, in contrast, was emotionally tiring labour, of unknowable duration. The responsibility of care within families in Japan typically falls on the eldest son or, more accurately, on his wife as described in Chapter 2. There is a feeling that sharing the emotional burden of care with family members would only make the situation worse, as Noriko-san explained to me: 'If you speak with your family about important matters, it gets more and more serious, darker…'

How then, do middle-aged carers receive support and an emotional outlet? One answer lies in typically all-female friendship groups, often developed in mothers' groups, workplaces or hobby groups; these can continue for decades, even after the original shared activity has long ceased. As documented in Chapter 4, through girls' nights (*joshikai*) and lunches, friendship groups are maintained by regular contact, now moving online to LINE chat groups to organise meet-ups and trips. The smartphone offers a lens through which the physical and social constraints of ageing and the burden of care can sometimes be mitigated.

> If I have a particularly hard day with my mother, my smartphone gives me a quick window to reach out to my friends and get sympathy. Being able to reach out to someone right at that second when you need them is the best thing about smartphones, and receiving stickers that tell me 'it's OK' is great.

The relationship a person has *with* their phone is deeply connected to the relationships that a person has *through* their phone, both to others and to themselves. Yet asking people about their relationship with their phone often yields limited responses: 'It's convenient for staying in touch' or 'I rely on it for everything'. People take for granted that the smartphone is a helpful tool, but they have typically not considered how it affects their relationships, behaviour and identity. The sparse response is both a methodological challenge and an ideal opportunity to bring together several of the findings discussed in this chapter and put them to practical use. If the visual has been so creatively appropriated as part of people's relationships with each other, it follows that through visual methods the relationship between ethnographer and research participant might be similarly enhanced.

I used graphic methods to explore the affective capacities of the smartphone in an attempt to make the ephemeral tangible. I asked Kazuko-san's group of friends to make a two-minute sketch of their relationship with their smartphone, to bring to the next lunch date. One of the members of the friendship group, Hiromi-san, had been undergoing chemotherapy for the past six months and she produced the most striking drawing out of the group (Fig. 5.6). It portrayed her at the centre holding her smartphone, surrounded by self-portraits displaying the range of ways in which she is affected by the smartphone in her daily life. During this time of illness and potential loneliness, the smartphone offers an escape from her present situation to her social relationships and the world beyond:

> Especially while I have been sick, the smartphone has become very important to me. It is my connection to the outside world. The days following chemotherapy my body feels drained and I cannot leave the house. During that time if I receive a LINE message or sticker from my friend I feel uplifted. But I can also feel sad and disappointed if I hear from my daughter that she is having relationship problems. When I am at the hospital having chemotherapy I watch films on Netflix and they often make me feel emotional. I also sometimes read surprising news stories. My smartphone makes me feel all of these things!

Through the smartphone, Hiromi-san experiences various affects and emotions because of her connections. Is it really the case, however, that the smartphone itself is making her feel these things? While the answer might appear to be an obvious 'no', there is more to the question than meets the eye. Not only is the smartphone her connection to the outside world, as she states, but it might also be playing a part in her connection to her feelings as well. In her drawings of affective states, such as surprise or happiness, she relies on the tropes of manga-style imagery and symbols. This prompts the question: are the aesthetics of manga which so shape LINE stickers also shaping the way in which people think about and represent their own feelings? Stickers express feelings, but in some ways are feelings also expressive of stickers, both reinforcing one another? As manga-like stickers become part of everyday communication in Japan and elsewhere, the influence of manga aesthetics may not only be facilitating the expression of emotions but also potentially shaping individuals' perceptions of their own feelings, creating a dynamic interplay between the medium and the emotions it conveys.

Figure 5.6 Self-portrait with smartphone. The text reads: 'The smartphone connects me to the outside world. So the smartphone has become a part of my daily life'. Drawing by anonymous research participant.

In questioning which emotions and affects can be rendered viable in a drawing or a sticker, one must also ask which cannot – what remains invisible and thus potentially unknowable, even to ourselves? Rather than simply documenting, images 'do things to us, interact with us, guide us, merge with us; they are literally part of our bodies and life-worlds', as the anthropologist Favero has put it.[71] There have been recent calls in anthropology to turn towards aesthetic modes of engaging with the world.[72] This is particularly important with digital media research because of the way in which the visual has become woven into the fabric of our relationships with each other. The book *Imagistic Care* presents a series of reflections from anthropologists, philosophers and artists on the relationship between the image and care, calling into question how images can fix or freeze perceptions, for example of old age.[73] Is it possible that the conventions of LINE stickers and other visual digital media are in some way reifying affective experiences?

The drawings by other members of the group were equally striking, but for different reasons. The portraits all showed the smartphone as being a focal point for attention, requiring two hands to hold it and an intent gaze and a seated posture (Fig. 5.7). When discussed with the group, this observation led to an understanding of the smartphone as being at the centre (*chuushin*) of their life and friendships; some participants had even written the word on their drawings. They agreed that the smartphone is an object not only physically close to them but emotionally central too, since it connects them with many of the important people and things in their lives. They would spend time at the end of

Figure 5.7 Sketches and texts showing people's relationships with their smartphones. Words and images by anonymous research participants.

their day on their smartphone, replying to messages and dedicating time to their friendships in preference to messaging on the go. For many of the middle-aged people in my research, beyond this friendship group, shifting from *garakei* to smartphones meant an increase in dependency on the device for daily activities. This increased ambivalence about how much time they spent on the device, but also made it an object they always wanted close to them.

Proximity is also important when thinking about how the smartphone is situated in a wider 'screen ecology' with other devices such as personal computers (*pasokon*) and tablets. Some people preferred to compose emails or long messages on their computers, which they felt were easier to type on than smartphones. However, for many older research participants the smartphones was their point of entry to the internet; they did not use computers at home. If people did have a range of devices, they would divide up their tasks based on ease of use. The smartphone was generally deemed to be the most convenient out of all their devices because of its portable nature. This has particular benefits for reading, as Sato-san, a flower arranging teacher in her nineties in Kyoto, explained:

> On a PC I have to lean in to look at the screen. But on this [the smartphone], it's so easy and I can make the screen and words bigger!

It was not only devices that were chosen for their suitability for certain tasks, but also applications and services. Emailing remained integral to how many people communicated, even with the rise of mobile messaging. People would use emails for communicating with friends and family, sending photos and for exchanging 'important' messages. Messaging on the smartphone was convenient for informal regular messaging, but email was preferred if they wanted to send something important. Sato-san explained:

> Email is more for longer sentences and things I would want to keep. Promises and things. Like, I ordered this and this is how many I ordered, things like that. So I can take a look later.

However, seeing the 'read' notification on LINE was reassuring for Sato-san. If she sent an important email she would follow it up with a message, just so that she could know that the person had received her communication. These findings regarding the 'screen ecologies' that emerge through personal choice and accessibility resonate with wider findings from the

ASSA project across our fieldsites.[74] In so doing they demonstrate the importance of studying the smartphone as situated alongside a range of technologies.

Gendered emotion work

It is important to consider not only how the smartphone is situated alongside other technologies, but also its place within normative social frameworks. For women, the smartphone emerged as especially important for a variety of reasons. Crucially, it has become a tool for female-to-female peer support as seen in the previous section. It fulfils a need for empathy and emotional support (*sasae*) which is sometimes lacking from husbands, as one woman explained:

> The support my husband gives me is different from my girlfriends because, women to women, they know what to say and the things you want to hear. I feel more understood when speaking with my friends. Men are more logical. I am not seeking for, you know, an answer. I just want someone to listen to me.

In central Kyoto, as mentioned in Chapter 2, neighbours typically know each other by sight but do not spend much time with one another. In my rural fieldsite in Kōchi, despite the town's small population, people often live geographically distant from friends, family and even neighbours. In both contexts, women often emphasised the importance of staying connected to a support network of friends through their smartphone. However, the smartphone is typically seen as a tool for keeping offline friendships going and for organising offline meet-ups rather than having friendships that are purely online. As one woman commented: 'My smartphone itself is not a cure for loneliness. It is seeing people every day that makes me feel better'.

As previously discussed in Chapter 2, among the older adults I worked with it was more commonly women rather than men who stayed abreast of the latest developments in mobile phone technology. They were more likely to have replaced their *garakei* with smartphones, primarily because their friends had done so and they wished to stay in touch through social media. Women tended to be more active than men in maintaining digital communication with friends and family who may be living geographically remotely, perpetuating the gender-based care roles discussed in Chapter 2.

It was also women who were more likely to attend smartphone classes. Their partners rely on them to teach them later or simply to show them the messages and photos that they receive on the messaging app LINE, so that they are not left out of family communication, as we saw in Chapter 2. My male research participants were more likely to reject the smartphone entirely, often because of cost and practicality. They felt that their *garakei*, or even their landline, still served them well enough. In rural Kōchi men would typically see friends and acquaintances when drinking at a local bar (*izakaya*); they would use their mobile phone minimally, arranging such meetings with either a quick voice call or a short message. In the cases where older men did own smartphones, they often protested that they did not know how to use them, or that they only used their smartphone for playing games and kept their *garakei* for making calls for reasons of cost.

That is not to say that men did not engage in affective communicative practices and performing 'emotion work'. Kenji-san, a man in his sixties who lived in central Kyoto, offers a different perspective on the affective nature of smartphones. When discussing why he went to great lengths to maintain communication with his friends by letter, Kenji-san told me

> Letters have heart (*tegami-ha kokoro-ga komotteiru*). Email has no heart. You can copy and paste your message on your phone and computer and it means nothing. Writing by hand means you care.

Kenji-san proudly displays letters and postcards he receives from friends, strung up against walls and collected in letter stands in his LDK (Living-dining room-kitchen). He showed me a pile of Christmas cards ready to be sent to his foreign friends, in addition to the 300 New Year cards (*nengajo*) that would soon go out – each with a hand-written message on the back. He keeps the old *nengajo* that he has received in a box. Each year's cards are carefully tied up in bundles, the oldest dating back 40 years. For Kenji-san, writing by hand shows deep feeling because of the effort required; it is certainly preferable to communicating online. His view of the smartphone as mediating cold communication, despite being completely opposed to the views of the women expressed above, ultimately supports the idea of the smartphone as an affective device which can mediate warm or cool relationships.

Digital distance

I have thus far discussed the affective affordances of digital visual communication as being central to how people maintain closeness and proximity. However, there is also another side to proximity which is equally important in relationships of care: distance. Kioko-san is in her fifties. She receives lots of messages from her father, who lives in a different city, throughout the day:

> My father has a smartphone and he sends me messages all the time, so many of them! Because it is so easy to send messages he tells me what he is eating and what he is doing. Giving him a smartphone is a way that I can care for him when I am not physically there. Although I feel he sends too many messages, it is easy to reply to him with a sticker.

This statement reveals the affective capacity of visual digital communication to enable frequent care at a distance, focused more on gesture rather than on content. Several research participants shared a similar sentiment. They felt that sending and receiving LINE stickers made frequent contact between parents and their adult children more likely and less burdensome than frequent phone calls in which one might struggle to find things to talk about, as anthropologist Tanja Ahlin observed in her work among Indian transnational 'care collectives'.[75] In Ahlin's ethnography among geographically distant family members, care was often communicated through the gesture of frequent calling and the sharing of mundane details from daily life. Webcams were also employed to enable a togetherness that did not rely on having things to talk about. However, Ahlin also highlights the practical care and observances of health status that are made possible through frequent contact, alongside the gesture of care.

The move towards frequent messaging prompts the question of whether it might also facilitate a withdrawal from phone calls, and indeed from face-to-face interaction more generally. People of all ages in this research tended to agree that face-to-face contact was preferable for communicating feeling and care. However, this just was not possible at the desired frequency given how geographically dispersed people often were from their parents, and how employment often left little time for anything else. This observation highlights the delicate balance between the convenience of digital communication and the irreplaceable value of face-to-face personal connection that many people were striving to get right.

In both Kyoto and Tosa-chō, middle-aged children were equipping their parents with smartphones and tablets to enable them to maintain frequent contact despite often living far away. However, even for people who lived geographically close to their parents, smartphones were being employed to insert a degree of distance, enabling their relationships to be better managed. As we saw with the intercom in the multigenerational home featured in Chapter 2, technologies are being used to mitigate over-closeness while still maintain connection. It turns out that anthropologist Jeannette Pols's phrase 'care at a distance'[76] can have multiple inflections when thinking about the role of the smartphone in informal care. I argue that the smartphone, in enabling 'care at a distance', explicates the case that sometimes care requires distance. According to Anne Marie Mol et al.,[77] 'good care' can be approached as 'persistent tinkering in a world full of ambivalences'. As seen in the case studies above, care is emotionally demanding. The smartphone is emerging as a key tool in the performance of emotion work to meet the demand.

In a wider context of an ageing society and concern over reducing burden on others, the smartphone can be understood as facilitating autonomy through enabling care at a distance. Many older people in my research valued being able to look things up for themselves on their smartphone, rather than having to bother relatives with constant phone calls when they needed something. However, it is important to consider the negative implications of people framing their technology adoption against feelings of burden, especially as technologies are increasingly being developed to aid elder care. While people in general desired their autonomy, they also wanted to feel that there was a safety net of caring humans with whom they would come into face-to-face contact.

It was older women in particular who appeared to value highly the autonomy afforded to them by the smartphone. When Junko-san, a 65-year-old woman in Kyoto, heard about the research I was conducting, she grabbed my arm and declared animatedly, 'My smartphone completely changed my life!' After her husband died, she suddenly realised how much she had relied on him for everyday things. He was the one who organised trips and drove them from place to place. Her daughter and grandchildren live in Tokyo, and it was the smartphone that came to the rescue when Junko-san had to start making this trip alone. Her daughter sent her train tickets via LINE message and she used her maps app to navigate the new city and public transport. At the time of my fieldwork, Junko-san had by now become adept at using her smartphone, but was frustrated with how un-user-friendly many websites were. When looking for details about a local festival one day, she showed me a long list of events that she needed to scroll through to find the right piece of information on a city website. '*This* is the digital divide!' she observed, meaning that her expectations did not often match up to the realities of poor web design.

Since getting her smartphone after her husband's death, Junko-san explained that she felt more connected with her daughter in Tokyo because of their frequent digital exchanges via LINE. However, she also appreciates the autonomy that the smartphone facilitates. Junko-san is a fan of watching figure skating videos on YouTube; via her iPad or Android smartphone she particularly likes watching male figure skating to 'gain energy' when she is feeling in need of a boost. She is aware that some female fans can be obsessive in their passion for particular figure skaters, so she only follows figure skating accounts on her anonymous and private Twitter account, not linked to any of her other social media. She does not want her friends and her daughter to see that she follows such accounts, so would never dream of following such content on Facebook where she is socially connected.

Junko-san's tactical use of technology to maintain distance and even invisibility from those who were closest to her was also evident in her use of visual digital communication. She would send stickers to her friends and was fond of emojis. She even categorised her phone contacts with emojis next to their names to remind her of how they met, such as flags for people she met abroad and houses to represent neighbours. However, she never used cute (*kawaii*) stickers with her daughter, feeling that they were too childish for her to send in this context.

In this mother-daughter relationship, Junko-san also valued distance in her offline interactions. She preferred to visit her daughter and grandchildren at their home in Tokyo rather than have them come to stay with her in Kyoto because otherwise she would have to tidy up all the small, non-child friendly objects that filled her home. Anthropologist Iza Kavedžija argues, in reference to older adults in Japan, that it is in keeping one's distance that it becomes possible to develop multiple dependencies.[78] I add that the smartphone is emerging as a key tool in this careful balance of autonomy and dependency that is at the heart of care practices between generations.

Conclusion

One of the primary contributions of digital anthropology has been to offer a counterbalance to computer, internet and technology studies that often start from the devices themselves rather than the people who use them. The title of this chapter refers to smartphones, but, as the ASSA project of which this research was part has shown, the meaning of the term 'smartphone' is always contextual.[79] To understand the smartphone in Japan one must acknowledge the lineage from previous technologies such as *garakei*. It then becomes equally important to acknowledge the dominance of one specific app, namely LINE. In turn, while LINE can be used in a multitude of ways, a primary affordance is messages and the sharing of specific genres of visual media. Within digital anthropology there is no such thing as 'the internet' or 'the smartphone', only a particular configuration of uses that we encounter through ethnography.[80]

I argue that the smartphone is an affective object, in the sense that it gives rise to various affects (such as anxiety, reassurance, care and disconnection) and mediates affective states in the same way that Yuriko-san explained how feelings are mediated by the 'atmosphere'. I argue that it is through visual digital communication that affect is most efficiently and commonly communicated via the smartphone by older adults. In creating digital 'atmosphere', stickers facilitate a sense of embodied proximity through which care can be realised.[81] I contend that a digital visual aesthetics of care is emerging that affords affective labour,[82] giving rise to a range of feelings such as a sense of support, as we saw in the friendship group above. In their communicative efficiency, stickers can also reduce the 'emotion work'[83] required in relationships of care. However, they were never considered to be a replacement for face-to-face contact and this was still deemed preferable by most people in terms of a hierarchy of ways to show and experience care. In contemporary Japan, where people may not live physically close to family or friends, care at a distance that carries some of the 'atmosphere', and thus proximity, of offline interaction is facilitated through visual digital media and is thus highly valued.

People often said that using visual digital communication brought them closer to others, making an analogy between physical and affective proximity. In engaging my participants in participatory drawing, I came closer to understanding their affective experiences. The anthropologist Rudi Colloredo-Mansfeld also found this to be the case in his use of drawing mediated encounters through which he 'began to see and feel in practical ways more of what they saw in their paintings'.[84] However, as anthropologists Azevedo and Ramos note, drawing is not a magic wand for anthropologists that allows us to see through the eyes of another; it is rather by trying to use 'other people's styles, observational techniques and aesthetics'[85] that we may get closer to them. I found that people were also better able to get closer to someone, and subsequently discuss their feelings, when using such an elicitation device. This was especially the case for topics that are hard to describe in words, such as public facade (*tatemae*) and true voice (*honne*). By spending time reflecting on and visualising their thoughts and feelings, for example towards their smartphone, I was able to elicit deeper considerations towards the device than through object-based elicitation using the smartphone alone.

Finally, this chapter has revealed the gendered usage of the smartphone, with women more readily embracing it than men. There is, of course, variation within gendered appraisals of the smartphone. I found that often this difference reflected the length of time for which a person had been using their device. Women relatively new to the smartphone were often highly enthusiastic about its capacities, whereas those who had already been using their device for several years revealed a sense of ambivalence about the degree to which it was now the centre (*chuushin*) of their lives. One woman in her sixties regretted that she now spends so much time on her smartphone, her household chores are a little 'less well done'.

Nevertheless, this chapter aimed to demonstrate how emerging digital repertoires have become so central to people's lives by in some way mitigating their sense of vulnerability and dependency on others as ageing persons, while simultaneously embedding them in relationships of care. In practice, I found women take on much responsibility for care within family and social relationships, making use of the visual affordances of digital communication to maintain warm contact, even when physically distant. Communicating affect through visual messaging while maintaining physical distance, the smartphone not only enables 'care at a distance',[86] but also demonstrates that in many cases care requires distance, and even tactical invisibility (as introduced in Chapter 4). The digital repertoires of older adults reveal how keeping one's distance can actually be what makes people feel close.

Notes

1. Stevenson 2014, 8.
2. Hochschild 1979.
3. Hänninen et al. 2021.
4. Hendry 1993.
5. Goodwin 2015.
6. Kavedžija 2018.
7. Thomas and Correa 2015.
8. Hardt and Negri 2000.
9. Hoschchild 1979.
10. Pols 2012.
11. Hänninen et al. 2021.
12. Hasebrink and Domeyer 2012.
13. Schwarzenegger 2020.
14. Kim 2017; Ito et al. 2005.
15. Ito 2005.
16. Kim 2017.

17. Kim 2017.
18. Kim 2017; Ito et al. 2005.
19. Okada 2005.
20. Kim 2017.
21. Japanese Ministry of Public Management, Home Affairs, Posts and Telecommunications 2002, as cited in Kim 2017.
22. Kim 2017.
23. Ito et al. 2005.
24. Ito 2005.
25. Hjorth and Kim 2011.
26. For a discussion of my remote video and graphic methods with Maiko-san, see Haapio-Kirk, forthcoming.
27. Mobile Society Research Institute 2021.
28. In my survey of 146 people aged between 40 and 60 years old living in Kyoto and Kōchi Prefectures, 100 per cent of respondents used a smartphone and 20 per cent also used a feature phone (*garakei*), primarily for phone calls. The popularity of the messaging app LINE among Japanese older people is high, with 76 per cent of respondents listing LINE as among their most-used apps.
29. Horst and Miller 2006.
30. McKay 2016; Wang 2016; Miller et al. 2016.
31. Cruz and Lehmuskallio 2016; Kress 2003; Sinanan et al. 2018.
32. Miller and Sinanan 2014; Jurgenson 2019.
33. Kress 2003.
34. Hjorth and Kim 2011.
35. Akimoto 2013.
36. Smith 2022.
37. Russell 2019.
38. Plath 1990, 231.
39. Steinberg 2020.
40. Kanayama 2003.
41. Negishi 2014.
42. Taylor and Taylor 1995.
43. Linecorp 2019.
44. Luor et al. 2010; Stark and Crawford 2015; Riordan 2017.
45. Lo 2008.
46. Dresner and Herring 2010.
47. Cooper 2018.
48. Malinowski 1923.
49. Sinanan 2019; Villi 2012.
50. Hall 1989, 91.
51. Nelson 2004.
52. Weicht 2013.
53. Fealy et al. 2012.
54. Loos and Ivan 2018, 164.
55. Minichiello et al. 2000.
56. Riach 2007.
57. Turkle 2011.
58. Kondo 1990.
59. Miller et al. 2016; Miller and Sinanan 2014.
60. Wang 2016.
61. Swift 2017, 152.
62. Lebra 2004.
63. Goodman and Refsing 1992, 3.
64. Befu 2002.
65. Hendry 1993.
66. Goffman 1959.
67. Hochschild 1979.
68. Cook 2013.
69. Tu and Du 1996.

70. Kavedžija 2018.
71. Favero 2018, 113.
72. See, for example, Pandian 2019, Ingold 2019, Grøn and Mattingly 2022.
73. Grøn and Mattingly 2022, 5.
74. See Miller et al. 2021, 65–70.
75. Ahlin 2020.
76. Pols 2012.
77. Mol et al. 2010, 7.
78. Kavedžija 2021.
79. Miller et al. 2021.
80. Horst and Miller 2020.
81. Goodwin 2015.
82. Hardt and Negri 2001.
83. Hoschchild 1979.
84. Colloredo-Mansfeld 2011, 5.
85. Azevedo and Ramos 2016, 146.
86. Mol et al. 2010.

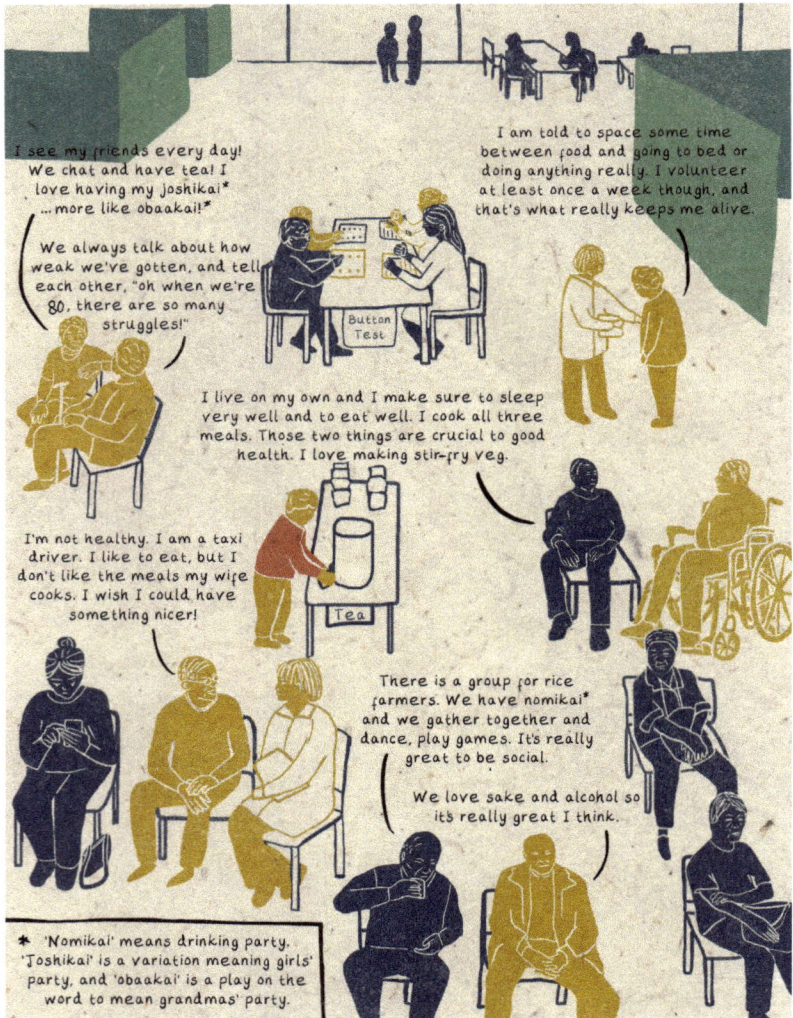

Image by Laura Haapio-Kirk.

6
Health: self-tracking and warm contact

> The government should put in a lot of money to prevent diseases. They don't do this. They know it is important, but they don't do it. There is not any progress on this. So we do not depend on the government. I depend on myself, it [health] is my responsibility.
>
> Kenji-san, aged 68, Kyoto.

Kenji-san is in his late sixties and lives in Kyoto with his sister, who is in her mid-seventies. Neither of them has children. He takes daily walks around Kyoto Imperial Palace for his health. In order to avoid ill-health he 'depends on himself'. To do so he takes advantage of the methods of self-tracking available to him, such as blood pressure monitoring at his gym, step counting on his feature phone (*garakei*), regular health checks and annual procedures such as endoscopy to check his gastrointestinal tract. Since his sister is older than him and they have no extended family, he feels the responsibility to care for her, should the need arise, is solely his. By engaging in various forms of self-monitoring, he feels better prepared to perform long-term care. While Kenji-san sees his self-tracking as a way to rely on himself for health rather than the state, in this chapter I broaden the definition of self-tracking beyond personal digital devices and individual tracking.[1] I explore the historical context of self-tracking and its current role in health checks and state-backed community care in an era of welfare reforms.

The dual decline of filial care for the elderly by the younger generation and of care from the state puts pressure on older people to act as carers themselves. Kenji-san does everything within his control to maintain his health, but he is particularly afraid of what is beyond his control – in particular of dementia (*boke*). He half-jokes that if he does develop symptoms he would commit a small crime, as he has heard that prisons have very good dementia care. In 2019 NHK, the national broadcaster, reported on the building of specialist facilities, akin to nursing homes, to care for elderly prisoners who make up a growing proportion of Japan's prison population.[2] Kenji-san, lacking the safety net of a family to care for him, believes that his uncertain future is compounded by the lack of care he feels from the government. As he observes,

> The government needs to be responsible for taking care of their people. To some extent they say they are working, but year by year we are becoming poorer. Life is very hard.

Neoliberal welfare reforms have influenced Kenji-san's feelings of uncertainty about more than his own future. He also worries about the financial and personal burden of being an elderly person caring for his elderly sister and how this will impede on his personal freedom to pursue his hobbies.

> As long as my sister lives it's okay. She is older than me, but I pay much attention to my health and she doesn't. No exercise! If she stays healthy I can do what I want to do. I can travel across Japan, but if she gets sick and I have to help her… it will be difficult.

His self-care activities described above are also an expression of care towards his sister, but he resents the fact that she has not been so considerate in her health choices, preferring to spend time in *pachinko* parlours[3] than at the gym. Both brother and sister enjoy pensions that allow them to follow lifestyles of their choosing, and their urban home facilitates easy access to services and their hobbies. However, Kenji-san feels that older people are paying the price of the current economic depression in Japan because of the care burdens placed upon individuals. Caring for his sister as a single man without additional support was a real concern, despite living in a densely populated urban area. This chapter will primarily focus on my rural fieldsite of Tosa-chō which, paradoxically, is revealed as a site of thriving community welfare initiatives, catalysed by the dual crisis of depopulation and super-ageing. In the last few years during the Covid-19 pandemic the inability of neoliberal capitalist governments around the world to meet welfare needs has demonstrated how conditions of crisis can promote anti-authoritarian forms of care, often operating outside of privatisation and austerity.

While other governments around the world were cutting social spending in the 1990s, Japan sought to mitigate some of the problems associated with a severely ageing population through developing social services for child-rearing and care for the elderly.[4] The establishment of the Long Term Care Insurance (LTCI) system in 2000 aimed to transfer the responsibility of elderly primary care from the family to the state, providing universal long-term care for every person aged over 65 under the slogan 'from care by family to care by society'.[5] Yet the burden of the latest economic recession has brought about parallels with other instances of neoliberal reform around the world. For example, since 2012 the so-called 'Abenomics' agenda of the Abe government has prioritised production and economic recovery over welfare and social protection.[6] While the LTCI system is still in place, the reality of an overstretched system 20 years later, with far higher numbers of elderly people needing care than when it began, means that families and communities shoulder the burden of care and can struggle to access the assistance they need. Often those assuming the brunt of care work in families are women,[7] though there are of course exceptions, as in the case of Kenji-san.

Kenji-san is an eager adopter of some forms of digital tracking, though he is reluctant to switch from his *garakei* to a smartphone, feeling that he can do everything he wants with it already. 'Digital health' now consists of a vast array of technologies ranging from health information websites to biometric monitors to bespoke smartphone applications. The World Health Organization defines 'mHealth' or 'mobile health' as 'medical and public health practice supported by mobile devices, such as mobile phones, patient monitoring devices, personal digital assistants (PDAs), and other wireless devices'.[8] The withdrawal of the welfare state in people's lives has arguably been exacerbated by digital health through its emphasis on individual responsibility for managing the body.[9] While such critiques offer essential divergence from optimistic 'techno-utopian' claims about the digital health revolution,[10] by focusing upon structures of power they may have neglected the ways in which people appropriate such technologies into their everyday lives or not, as the case may be.[11]

In the Japanese context, where the practice of self-tracking for health has a long history pre-dating the current digital 'revolution', a broader and more holistic view of self-tracking emerges. In this chapter I aim to reveal a historic understanding of the wide range of corporate and government-run services which are responsible for the health and welfare of people. Through ethnographic stories of individuals engaging in various forms of self-care, I aim to show how health tracking practices developed before the rise of digital tracking devices, involving a network of health professionals, family, community, and even spiritual entities. Rather than being an individualising or self-interested practice, as has been a prominent critique,[12] health tracking can be highly social and involve a balancing act of interdependences and mutual contact.

This chapter will suggest that the concept of self-tracking can be expanded to include forms of data not easily quantifiable, such as visual content shared by older users. This approach serves to build on a growing area of anthropological research on ageing and technology that seeks to rethink 'age through technology while rethinking technology through age'.[13] By studying how older people are tracking and maintaining their health, both digitally and offline, I argue for an expansion of the scope of mHealth to be more inclusive, accessible and embedded within already existing digital and social repertoires.[14] By doing so, it has the potential to go beyond the self-knowledge of biometric data to exploit the inherently social nature of health. Building on the concept of mutual assistance (*yui*), discussed in Chapter 4, this chapter shows how such ideas are foundational to preventative health initiatives both online and offline.

The burden of care

OECD[15] countries share a common demographic trend of having shrinking and ageing populations, but this trend is the most pronounced in Japan. It has the oldest population in the world[16] and it is in rural parts of the country that the population is ageing fastest. The percentage of the population of my rural fieldsite of Tosa-chō aged over 65 is almost double the national average, at 45 per cent.[17] Associated with the ageing population is the wider context of severe depopulation that has occurred in rural areas across Japan. Unlike Kyoto-based Kenji-san and his sister discussed above, older adults in rural areas are often unable to access amenities easily. When living alone in depopulating regions, social isolation becomes a prevalent problem. This in turn presents a risk factor for health, including being associated with poor nutrition,[18] cardiovascular disease,[19] reduced cognition,[20] mental health problems and Alzheimer's disease[21] and increased mortality.[22]

Despite the issues associated with rural depopulation, recent ethnographic research has found that conditions for older people in Japan can be favourable for wellbeing in rural areas.[23] The number of unattended or solitary deaths among the elderly in urban communities outnumbers those in rural areas, suggesting that social ties remain stronger in the latter.[24] This argument accords with my findings about the persistence of relationships of mutual support (*yui*), as discussed in Chapter 4, especially in rural areas. This is a key motivation behind migration from cities to the countryside, as will be discussed in Chapter 7.

Since 1965, the population of Tosa-chō has shrunk by 50 per cent to 3,807 people as of the end of March 2020.[25] As one local man in his seventies explained, 'there are more and more houses without lights on'. He feared that having fewer neighbours will make life harder for elderly residents in a community traditionally sustained by mutual forms of care and assistance. Professional home care was perceived as being limited in remote regions with declining populations and less viable for people with lower incomes compared with those living in cities. This situation of reduced access to professional care combined with weakened local communities has implications especially for older adults who live alone – which, in Tosa-chō at the time of my fieldwork, comprised 256 people over the age of 65 and 70 people over the age of 75.[26] In response, the local Social Welfare Council (*Shakai-Fukushi-Kyogikai* or *Shakyō* for short) had developed a 'watch over' system by which they could track potentially vulnerable older adults through visits by volunteers and professionals such as postal workers.

I accompanied a volunteer on one such visit to a widower in his eighties who lived near the top of a mountain. The volunteer explained that, although they were not related, this man felt like an uncle as they had known each other for 40 years. The older man had helped him when he had first moved to the region from a neighbouring prefecture. As such, the relationship was more one of reciprocal care and trust built over a lifetime than a one-directional performance of care, as is more common in professional care work.

The visit was a chance to chat and check the general welfare of this older man, rather than performing specific care duties. Just as a son might do, the volunteer casually turned over a page on the calendar hanging on the wall to display the correct month – a minor act but one that signified genuine care and required physical presence. Also adorning the wall was a handbag which had belonged to the older man's wife. One imagines that she might previously have been the one to flip the page of the calendar, the small daily routines and chores into which one settles as a couple. The older man was strong and continued to chop firewood; he had stockpiled a large amount which lined the outer walls of his house. He would need the supply for the coming winter to heat his woodfired bath that stood in an outhouse.

Although this man was still physically strong, falls and other accidents could take on greater significance now that he lived alone. A sensor had been installed on the ceiling above his bed, linking him with a company that monitored for inactivity and would alert local welfare organisations if they detected unusual patterns. The man also had an alarm button on a lanyard, but he found it rather ineffectual and chose not to wear it. Once when he had fallen down he had pressed the button. However, because of the remote location of his house, it was a neighbour who came to the rescue before any official service.

The Social Welfare Council promoted community health through measures including a club for older adults and events targeting depopulated villages, often held at one of the nine abandoned schools in the region. These schools have been repurposed as 'warm connection centres' (*attaka furei sentā*), in which residents are encouraged to organise their own activities, such as a 'warm festival' (*attaka matsuri*) where people can sell their handicrafts and a 'warm holiday project' (*attaka yasumi purojekuto*) for schoolchildren to interact with local residents during the long vacation. Kōchi prefectural government has established 200 of these 'warm connection centres' in rural areas across the prefecture. The centres are designed as places for everyone, rather than being dedicated only for the elderly, for disabled people or for children, as the numbers for each individual group are small in semi-mountainous areas.[27]

Putting policy into practice, the Social Welfare Council worked to foster a community in which residents are supported in helping each other, just as they did through the old system of *yui*, now threatened by depopulation. As Yamakubi-san, who ran the Council, explained:

> What is most important to us is to bring the really rural communities and the lively ones together and make everyone happy. Creating a sense of revitalisation for everyone. When there is a village that is lacking support or is dying out, the Social Welfare Council will gather ideas on how we can revitalise that community, and those people there. And from there, we work with others … We are the cane that enables the dying community to stand on their own. We don't build the relationship, but we support the dying community to be able to build that relationship on their own.

In rural areas where older adults may be living far from family members, communities play an especially supportive role in welfare provision. The Social Welfare Office in Tosa-chō has grown in size and responsibility over the past 20 years, with volunteers assuming more of a central role in caring for the ageing community as support from local government decreased. Yamakubi-san explained:

I think the country [national government] realised it was difficult to tackle so many issues, so they decided to hand out those problems to each town and its local people. Also, many activities are much cheaper to be done by the Social Welfare Council (shakyō) rather than the Town Hall (yakuba).

The Social Welfare Council received funding from a range of sources, including through centralised government funding and from donations directly from local citizens. In this sense, it blurred the boundary between a state-supported and grass-roots organisation.

As anthropologist Nana Okura Gagné argues, Japanese citizens are familiar with marketisation and the idea that the state is not obligated for citizen welfare. The emergence of neoliberal rhetoric in post-bubble Japan has thus not elicited 'complete bewilderment nor unbridled enthusiasm'.[28] Rather, neoliberal reforms have prompted debate about the future of the corporate world, Gagné claims. Japan's particular history of corporate familism meant that one's company typically took responsibility for the health of employees and their families.[29] Rather than 'welfare statism', Japan promoted 'welfare corporatism',[30] looking after employees and their families often through annual company health checks. While corporate healthcare is diminishing as more people are in precarious forms of employment without benefits, the provision of health and welfare is simultaneously diminishing from the state.

This dual withdrawal could indicate that the responsibility for health is placed thus solely on the individual and the family. However, since 2015 the Japanese government have shifted the strategy of its Long-term Care Insurance Act towards community-based care and social determinants of health, such as with the promotion of community salons (ikoino saron) run for and by older adults themselves.[31] Such initiatives are common across Japan as a way of combating loneliness through mutual aid. Community salons tend to be run by residents and volunteers, sometimes supported by other relevant organisations such as the local Social Welfare Council, as private, non-profit-making operations.

In Tosa-chō the older adults' community club is called the House of Tonkararin (tonkara-rin-no-ie), playing on the lyrics of an old wartime song and bringing to mind the 'ton ton' noise of tapping on a neighbour's door or window. In his account of 'phatic traces', the linguistic anthropologist Nozawa recounts how the idiom of contact, or 'touching together' (fureai), has become intrinsically associated with elderly care and services in response to a society-wide concern over a loss of connection (muen).[32] Community clubs such as Tonkararin are known as *fureai ikiiki saron*, which translates as 'lively mutual contact salons'. The invocation of a friendly neighbourly tapping represents a nostalgia for a possibly mythical time in which community was abundant and neighbours helped one another in times of need, or were at least present to say hello to one another. Started in 2006 by local volunteers, Tonkararin has grown to be a key institution for the wellbeing of the area's older population.

With the decline of forestry and agriculture in Kōchi Prefecture, young people in rural areas soon left for cities. It is striking, therefore, that the club is housed in a disused museum of forestry and woodworking. The exercise class is held in the 'Forest Theatre', a large room with a stage flanked by trees with resident owls. After one lively exercise class one of the attendees, to prove her fitness and lung capacity, performed an impressively long 'cock-a-doodle-doo' (*kokekokkō*) to the applause of the club members. Upstairs, near a smaller community room and kitchen, the walls are adorned not only with images of trees and information about local types of wood, but also photographs of members of the club. One photo showed a couple in their eighties dressed up in Western-style wedding attire, marking the fulfilment of their wish to have such a celebration in later life.

The community salon connected people whose communities had been severely affected by depopulation caused by the loss of forestry and agriculture. A holistic approach to wellbeing, taking the importance of sociality for health seriously, informed the design of available activities. Open four days per week, and with a rotating list of attendees who would come from far and wide with provided transport, the club offered an exercise class, games and social activities, along with meals and snacks. The cost of attendance was 650 yen, or some people paid with rice (two *shō*, roughly 3.6 litres). Since 2016 transportation, previously covered by participation fees and donations, had been covered by government subsidies from Kōchi Prefecture Warm Contact Center Business Expenses (*Kōchi ken atta ka fureai sentā jigyō-hi*). The regularity of 'warm contact' through attendance at the club was seen to work as a preventative measure, offering a chance to catch health and welfare issues early. People who lived in remote areas were brought together and, through the mutual attentiveness and care of the volunteers, those who were in need of extra support were referred to the Social Welfare Council. The same idea of preventative medicine was behind the annual community health check, to which this chapter now turns. The 'warmth' of contact was central to the success of this annual check, which attendees greatly looked forward to as a social event.

The health check

Tracking health via large-scale community events has been a strategy central to healthcare provision in Japan since the post-war welfare reforms. Routine health checks are made available to almost all segments of population throughout the life course, including a specific health check for those aged between 40 and 74, introduced in 2008.[33] Health checks are typically organised by health insurance companies, of which there are over 3,000 in Japan; they are also provided by some municipalities and privately. The focus is on the prevention of lifestyle diseases through quantifiable measures and the encouragement of personal responsibility for health. Based on the results of various tests, including blood tests, urine tests and measuring weight etc., guidance is provided to individuals by doctors and nurses. Since 2017 tele-consultation has been allowed to improve the efficiency of providing health guidance.[34]

My research revealed that while the health check does indeed rely on the quantification of the body to prevent and decrease the effects of lifestyle diseases, the social experience and temporality of the health check as an event meant that people felt that they gained health benefits simply by attending. The annual health check in Tosa-chō at which I volunteered enmeshed a person in a practice of goal-setting dependent on promised future checks and the supervising gaze of health professionals. Rather than an individualising practice that creates self-knowledge, as in forms of digital self-tracking, health checks embed people in interdependent relationships with health professionals, and more broadly the state, often over long durations of time. They are also simply occasions to be social, a chance for participants to meet friends and acquaintances while waiting to be seen by doctors. Social activity was deemed important for health by many of the participants in the ethnography, as shown in the illustration before the start of this chapter. These observations accord with an emic understanding of health in Japan – that is, they align with the internal, culturally-specific perspectives and values regarding wellbeing within Japanese society that place interdependence and social relations in high regard.[35]

The annual health check for over 75s in Tosa-chō takes place over five days in August. It is held in a large building attached to the town hall. In the video below, social nutritionist Kimura Yumi talks about how this health check, in combination with wider community efforts, has helped to improve the longevity of the local population (Fig. 6.1).[36] At the 2018 health check, along with around 50 students and researchers from universities across Japan, I and my research assistant Sasaki Lise assisted

Figure 6.1 A short video about the annual health check in Tosa-chō for those aged over 75. https://youtu.be/N2UUht2jG7Q.

Kimura. I observed how people arrived individually in their own cars, or in groups, collected from their villages by minibuses that wind their way through rice terraces and around mountain roads to ensure that everyone who wanted to attend could do so even if they could not drive.

About 20 different stations were set up for various tests, including a physical test of co-ordination, a cognitive test, dental checks, measurements of body fat and muscle percentages, finger dexterity tests, a blood test, a driving simulation test and various others. People tended to be most anxious about taking the cognitive and driving tests, as these were seen to be the ones that could negatively affect their lives if they performed 'badly'. Otherwise the mood was jovial; during the waiting times, as people moved between the different stations, they sat and drank tea and chatted with friends and acquaintances. A tea station was prominently positioned in the centre of the room and blue screens separated the different stations, providing privacy.

This health check in Tosa-chō began in 2004 and the project is known among the doctors that run it as Field Medicine (*firudo igaku*).[37] The annual event allows the doctors to monitor the health of the elderly, research the mechanisms of ageing and investigate concepts such as 'healthy ageing' and 'preventative care' within the rural communities that are the face of 'super ageing' Japan. Kimura uses data collected at the health check and via a pre-circulated questionnaire to understand how nutritional choices affect physical and mental health status; she also studies how they are influenced by social factors such as eating alone.[38]

Measurements of body fat and muscle percentage allow her to track the effects of nutrition on the body over time to make evidence-based recommendations for elderly populations.

Kimura, along with the other doctors present at the health check, offered consultations for any patients whose measurements raised concern. Patients brought with them dedicated notebooks, given to them in previous years, in which the consulting doctor wrote their recommendations for how they should proceed over the course of the next year. The doctor doing the final consultation also took a photo of themselves with the patient and glued it into their book to help build up a visual record of the patient's experience at the health check. These selfie-like images showed informality and genuine connection. Some patients waited to see specific doctors with whom they had built a relationship over the years.

Associated with the health check were various occasions for the doctors and locals to mix in informal settings. One such event was a meal where members of the elderly club performed a dance, replete with bright pink sequinned costumes. Kimura was visibly moved when watching the dance and observing people whom she had known since the health check first began, and were thus now advanced in age, yet still able to dance and put on a great show. It was clear in this moment that the health check was about far more than just data for both the patients and doctors involved.

Sasaki and I were stationed on a machine that measured body fat and muscle. This positioning allowed us to talk with hundreds of patients about their lives and general attitudes towards health while placing electrodes on their ankles and fingers and waiting for the machine to take its measurements. People were generally interested to hear whether their measurements meant they were healthy, but the numbers themselves did not interest most people. What emerged from talking with people was the sense that attending the health check itself made them healthy, rather than the data gathered at the health check. The event supplemented other regular practices that helped them feel that they were positively influencing their health. Such practices mostly revolved around the daily production and consumption of food. Participation in agriculture was seen as healthy for the body because of the physical strength and vitality required, while the consumption of home-grown produce was considered optimal for health.

Attendance at the health check was a motivational practice; it encouraged many people to 'do their best' and follow their doctor's advice until the health check the following year. Sometimes this advice might be to aim to lower a certain reading, such as blood pressure. As Wada-san, a woman in her late eighties explained, goals are very important to staying healthy in later life:

> Thank you; I am healthy because of this check-up. This motivates me to be healthy. Goals so are important. In anything, like, I am going to accomplish something by tomorrow or by next month. Whatever it is, it's so important to have a goal. Staying at home and not doing anything is so bad – you have to get out of the house. I like to plan myself. When you get older, you are against nature. So you have to face it and acknowledge its existence. Live with it. I am over 80 now. In the summer, it is so hot. So I get up at 5 a.m., and I eat a salt candy and I walk for about 40 minutes, then I have breakfast. At 6:30 a.m. I lie down and I start my radio exercise. My doctor told me my blood pressure is high so I stopped putting too much sugar in my coffee. This is my body, and I need to take care of myself.

For Wada-san, taking care of herself meant a form of dependence on the encouragement and structure provided by the health check and more generally on her regular routine. Self-care through goal setting was developed within wider structures of institutional care.

What motivated Wada-san, like many of the people who attended the health check, was the desire to maintain her independence in later life. While an emphasis on independence aligns with the goals of neoliberal policy, in this case the motivation behind the striving for independence was highly social. It related to Wada-san's desire not to burden her family in her old age, a recurrent theme in the wider ethnography:

> Being independent will not burden anyone around me, especially my family. It is not good to burden my family for my health issues. So I stay active. I have been working until age 78 because I was working at a store. After that my husband died and then grandma [her mother-in-law] died and I finally realised I needed to take care of myself and be individual. I never thought that way when I was working, but I turned myself around … It has been ten years since my husband passed. When you are alone, you have to do things on your own. You have to take action.

Wada-san had dedicated her life to caring for others, yet she was reluctant to place the same expectation of care on younger generations. She was motivated to stay healthy for the sake of herself and her family, mainly through gardening and taking care of her family home, a short drive from where she lived. She thought of herself as the 'protector' of this home and garden, tending to it regularly. When we met, she proudly showed me photos that she had taken in her garden on her new smartphone. She mentioned that she had many more photos on her digital camera of the flowers which she felt so passionately about:

> So now, there are many, many flowers from plum to sakura and others. When the colours of the flowers and the leaves change, they fall to the pond underneath and are so beautiful.

Attending the health check contributed to Wada-san's sense that she was actively involved in maintaining her health. It was the practice of health tracking rather than the data produced that was most important to her feeling that she had agency in the management of health. Recently she had acquired a smartphone, but she was not interested in using it to track her health, nor did she feel the need to use it to access the internet: 'I have my local supermarket and communities so I would never use my smartphone for that. It will cost me too!' Many participants had mobile phones poking out of shirt pockets or in their hands as they waited to be seen by doctors. While it was clear that many people had feature phones (*garakei*), those who did have so-called 'easy (*raku-raku*) smartphones', designed for older users, were often highly enthusiastic about them and proud to say that they owned one.

In the preceding account of the rural health check and of the community club, two observations emerge. First, participation in various tests is less about gaining self-knowledge and is more to do with highly social reasons of wanting to do everything possible to stay healthy so as not to burden one's family in old age. The burden of care posed by the ageing population is a dominant concern in Japan, permeating people's attitudes towards themselves as ageing persons. Second, the health check and community club are primarily seen as social events, allowing attendees to see their friends who may be living remotely. This specific health check was also a social event for the doctors who attend from all over Japan and maintained long-term relationships developed with the community.

Based on these observations, especially the emphasis on commensality, the act of eating together, I subsequently collaborated with Sasaki and Kimura in creating our own small mHealth intervention which will be discussed below. This project – the 'smartphone food club', conducted with a small number of participants – encouraged healthy eating through digital social interaction. However, there are initiatives much bigger in scope, such as an official foray into mHealth by the local prefectural administration, to which I now turn.

Kōchi health passport

Kōchi Prefecture regularly ranks among the poorest of Japan's 47 prefectures.[39] It also has one of the highest rates of death among working-age people from preventable diseases, such as stroke and diabetes, in the country.[40] The associated high medical expenditure is one of the central motivations for the local government to focus on preventive action for non-communicable diseases, which require long-term continuous medical treatment and are thus more costly than acute diseases. The prefecture is famous for its gregarious inhabitants' love of sake and lifestyle diseases, such as cardiac disease and hypertension, are common.[41]

A team at the Kōchi Prefectural Office is dedicated to improving the longevity of its citizens, to bring the figures up to Japan's famously high standards. One of the latest strategies of the Health and Longevity team is a system called the 'health passport' that they started in 2016. 'Five stickers and you can enter to win a prize', the head of the scheme explained, showing a series of small, passport-style books with empty slots where stickers could be placed (Fig. 6.2). Once a person had filled up their passport with stickers they could proceed to the next colour and 'rank up'. The team leader continued:

> If you go to a (health) seminar or any workshop you get a sticker, and the more you get, you rank up. These health passports are beneficial because just by showing this at public facilities such as the gym or the supermarket, you can get a discount. At a restaurant, just by showing this you might get a free drink or a dessert ... If you rank up you can get better benefits.

She explained that their health passport app was only for Kōchi, but that all prefectures in Japan were making their own health apps as part of wider efforts to promote longevity.

Hikaru-san is a local Tosa-chō resident in his mid-forties. He used a different self-tracking app for a while to help him to lose weight, but explained that people in their sixties and seventies 'love points cards' and therefore would be interested in the gamified approach to mHealth that the Kōchi Prefecture app represents. He explained that the gamification of health through such a rewards system is not so far removed from

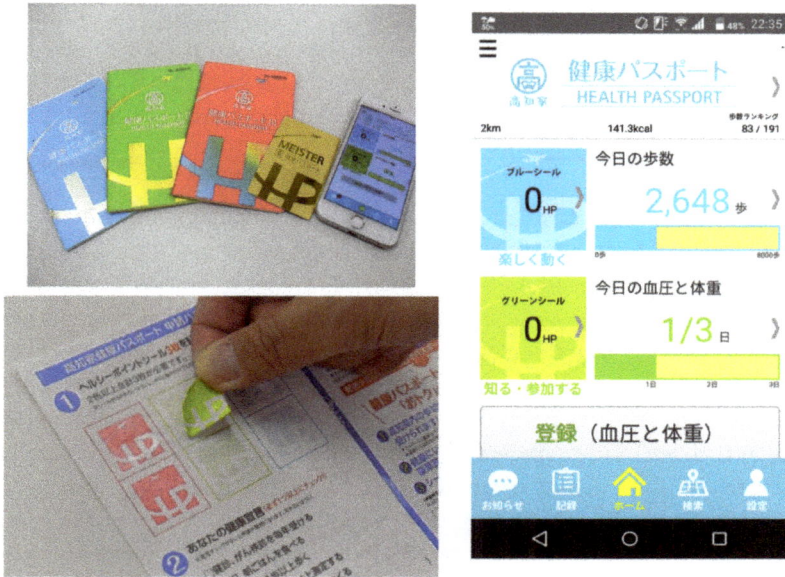

Figure 6.2   Physical and digital health passports. Images by Kōchi Prefectural Office.

the way in which pilgrims on Shikoku's famous pilgrimage trail rise up the ranks the more temples they visit.

> Kōchi has a culture of pilgrimage ... those people that go around the temples? Yeah, so that's more like a stamp rally and the more stamps you have, you gain a next level. So I think Kōchi people love that stuff, I think Kōchi people love collecting things.

Hikaru-san was referring to the 88 temples dedicated to Buddhist monk Kōbō Daishi that comprise the circular 1,200 km pilgrimage route which is still popular today. The more temples a pilgrim visits, the better the spiritual rewards. To mark the number of times a person completes the whole pilgrimage, they receive a different colour prayer slip (*ofuda*) to leave at each temple. On the note the pilgrim writes their name, address and prayer.

> There are so many! Red, white, yellow. The ones that are superior are silver and gold, and then I think the very last most superior one is *nishiki* – like a Japanese folded pattern one. I don't know if this is still a fact, but I know that before, if you had gone around the

temples more than 100 times, then you would receive this *nishiki* note. And this *nishiki* note was known to be good fortune. For example, if you have that one then all illness will be cured or your business will go well ...There was actually someone selling those notes! For 5,000 yen [£35] or something.

The link between spirituality and commodification has a long history, as the work of anthropologist Inge Daniels demonstrates. Daniels's most recent work on Japanese New Year cards shows how they play a vital role in nurturing social, spiritual and economic life in contemporary Japan, challenging the dichotomy between ritual and economic practices while proposing that cosmological beliefs may profoundly influence economic and political spheres.[42] She notes how this connection is at the root of the Japanese term for souvenir, *omiyage*, which literally means 'shrine box' and originally would have referred to gifts procured by pilgrims on their travels.[43] In the *Meiji-Taisho* era (1868–1925), capitalism was becoming foundational in Japanese society and many cultural and spiritual objects were becoming commodified.[44] Obtaining souvenirs at temples is still thought to bring luck or health. This practice is not so far removed from the health passport system of obtaining stickers at health venues to enable one to obtain rewards. In short, the connection between self-tracking and health, often with visual markers, goes back centuries in Japan.

Tracking progress via stamps and stickers in dedicated books is common for every kind of activity imaginable in Japan, from hiking to visiting castles, as mentioned in Chapter 2. It is no surprise then that the health passports proved popular among the people of rural Tosa-chō. Attendees at the health check described above received a passport sticker, for which people would queue up. There were few other opportunities in this rural town to collect stickers, with many health events taking place in the city and limited access to sports facilities that could reward exercise. The bias towards city-dwelling users was one of the reasons why the Kōchi Longevity team decided to create a smartphone application to accompany the health passport that would reward people with stickers as they carried out their daily lives in a healthy way.

If you input your weight three days in a row, you get one sticker. If you walk 8,000 steps you get one sticker.

The app was deemed to be also beneficial for busy working-age people who might not have time to attend a sports facility. The design of the app was not a case of the digital replacing the physical booklets – rather, the digital was designed to complement the sticker system. Once a user had achieved a certain number of stickers in the app, they needed to tap a button; this sent a message alerting the app team to send the user a physical sticker in the post. This could then be affixed to their physical passport. Even with the collection of digital stickers, users needed to record their progress in physical books. This integration of physical bureaucracy in supposedly digital systems is common in Japan. The practice was highlighted during the Covid-19 pandemic when people were advised to register for their benefit payment online. When people tried to reset their passwords, they were told to head to their local town hall to do it in person, resulting in widespread queues and derision across Japan.

The app did not replace the physical passports initially. It served only to increase a person's chances of receiving coveted stickers, yet usage among the older people of Tosa-chō was rare. However, after several years, the health passport went entirely online in 2022. This protracted move to a digital platform may be because, at the time of my ethnography, smartphones were only just starting to be used by older people in the community – largely due to the previously prohibitive cost of smartphone plans and the enduring popularity of feature phones (*garakei*), especially among older people. In Tosa-chō a smartphone club had been established in 2019 by the town hall. People could come to the club to learn how to use their smartphone by playing games and receiving advice from town hall staff and employees of the phone company Docomo.[45]

The kind of smartphones that many people had were what are known as '*raku-raku*' phones, or simple smartphones. These had bigger icons and interfaces and simple navigation. They were often available in plans advertised as being for the over-sixties which would cost between 3,000 and 4,000 yen (£20 to £30) and came with unlimited talk time and fairly low data, such as 1 gigabyte. As discussed in Chapter 5, the new adopters sometimes conflated using a smartphone entirely with the use of LINE.

A digital response to isolation

Given that traditions around reciprocity and sociality are hampered by physical distances in this mountainous region, Tosa-chō provided an ideal context for a digital health intervention facilitating sociality at a distance. Upon my return from fieldwork I collaborated with Sasaki and Kimura, introduced above, to devise a digital health intervention among older people in Tosa-chō. From my ethnography it was clear that it would have to be via LINE, and that it should draw as much as possible on already existing social practices of commensality and visual digital communication.

A full discussion of the resulting intervention – a food club established through sharing photos of meals on LINE – has been published elsewhere,[46] along with an account of the methods and results, including changes to the Quality Of Life (QOL) scores of participants during the intervention.[47] The aim of our intervention was to explore what social mHealth could look like for this population of recent smartphone adopters, and to mitigate some of the potential health risks of social isolation that face this depopulating community.

An important but under-explored area of mHealth is to reduce the risk of social isolation for this population by building on their already existing digital repertoires of visual messaging. While smartphones are only now starting to be used by the older population in Japan, studies have shown that digital technologies such as social media have the potential to enable greater opportunities for social connection for older adults, while improving quality of life and reducing loneliness.[48] Frequent use of social technologies has been associated in some studies with increased self-reported feelings of health and wellbeing and a reduction in depressive symptoms among older adults.[49] However, other studies have demonstrated no or limited linkage between social technology use and improved health and wellbeing,[50] so further research in this area is required.

Sharing photos and messages about food in a social chat setting is not the same as keeping a regular record of exact nutritional data, of course. Yet it served to cast a social gaze during mealtimes that would otherwise be lacking. This social context made people more aware of the food they were eating. They were encouraged by each other when sharing photos of home-grown produce and home-made meals (Fig. 6.3), so digital sharing may have influenced their nutritional choices. As Kimura observed at the start of the intervention, it is very difficult to ascertain correct nutritional information from photos – after all, who knows what is contained in a bowl of curry? The potential of a digital nutritional intervention based on photos of food is consequently limited. Yet the social aspect of eating is something that may have once been common in the lives of these now elderly participants, many of whom may now eat alone. Participants said that the intervention made them more inclined to try new recipes and to put more effort into their cooking. As one participant said, 'It gave me great inspiration and raised self-awareness that

"Today was a rainy day. Today's dinner was Japanese style with a lot of vegetables."

"Japanese Carrots!"

"Those are wonderful vegetables! Did you grow them?"

"Side dishes were mackerel, some marinated vegetables like daikon, shiitake and fried tofu. I also made some pickled Daikon with broccoli."

"Marinated vegetables are great as we age 😊"

"Daikon, chinese cabbage, mizuna, carrots and spring onions. I will bring them to my daughters place tomorrow."

Figure 6.3 Typical messages shared in the smartphone food club – a mixture of text, photos, emojis and stickers. Image by Laura Haapio-Kirk.

'I should do better!' Another observed: 'I'm happy. I'm expanding my repertoire. My eating habits are becoming richer.'

One of the strongest indicators of the smartphone's potential for welfare initiatives among older populations is that activity in the LINE food club did not cease once the intervention was over at the end of February 2020. The timing happened to coincide with the start of the Covid-19 pandemic in Japan. In addition to photos of food, participants continued to share messages and photos for the next 11 months. One of the critiques of mHealth is that usage can have limited longevity, with abandonment by users common.[51] As discussed earlier, this intervention built on emerging and pre-existing digital repertoires, namely using LINE for communicating with family. It was thus embedded within wider practices and did not require special knowledge or equipment.

In the early days of the Covid-19 pandemic participants shared information about how to stay safe, photos of face masks they had made for their grandchildren and other encouraging messages. Yamakubi-san, the head of the Social Welfare Council, sent informational messages. The chat group became a vital source of sociality in a context where face-to-face interaction was severely limited. Yamakubi-san, upon making this observation herself, then decided to replicate the intervention within the wider community by setting up a system of 'LINE buddies'. While the LINE buddy chat groups were not themed around food, as a reward for signing up to the scheme participants were promised a meal in the town hall together when restrictions eased. It seems that sharing meals both online and offline will continue to be central to sociality in this town. As many residents did not have smartphones, the scheme was also open to those with landlines. What mattered was maintaining connection. As one participant in the smartphone food club commented:

> When connections are lost, cultural traditions disappear. The fact that there are no longer families that live together for three generations means that Japan's history, its 'living history' (*ikite kita rekishi*), cannot be passed down. With the spread of smartphones, I feel that connections between people are increasing, but I also feel like some culture is disappearing. As a result, Japan's sharing culture (*osusowake bunka*) and way of thinking as a human being will disappear. That's because all the people who could pass on the message are gone and the way we live our lives is becoming more convenient.

For this woman, the rise of digital communication had the potential to strengthen ties between generations. However, she saw the smartphone as a double-edged sword, simultaneously making life more convenient and thus reducing people's need to rely on each other. What was at stake was the cultural tradition of sharing (*osusowake*) that define social relations in this rural area. Our intervention, along with the subsequent action taken by the Social Welfare Council, demonstrated the smartphone's potential to strengthen such fragile webs of sociality, particularly in times of crisis.

During the Covid-19 pandemic all face-to-face activities, including the community club Tonkararin, ceased. This pause was concerning for everyone involved. It echoed wider concerns I had heard about the future of the initiative, relying as it did on the work of around 70 volunteer staff, all over the age of 60. In an ageing and depopulating context, it was uncertain who would run such initiatives in the future, since many of the organisers were already in their late seventies and eighties.

One of the women who had been instrumental in setting up the club explained the importance for everyone to be involved in the running of it as volunteers:

> I stand for everyone's right to be individuals and do what they want to do, however they want to. Everyone has their unique characteristics – some like to cook, some are very good at crafts. We have meetings and see which person can be in charge of what.

During my fieldwork, a part-time management position was funded by the town hall to facilitate the smooth running of the club. With a move towards a more top-down running of the organisation, this egalitarian ethos, while still strong, was not guaranteed to remain into the future. The founding spirit of the club was that everyone could be a leader, as the volunteer continued:

> For me, it was really about creating a town of happiness. This is purpose in life (*ikigai*). This place is a space for dreams. It is where dreams come true and every day is more productive … This has also led to bringing everyone together. And by binding everyone together, they share the idea of helping each other … Everyone is a leader and everyone contributes to our society.

Tonkararin provides an opportunity for reciprocal care, for warm contact and for people to fulfil responsibilities and roles. As a side-effect, it also enables the informal tracking of health among geographically dispersed communities. It remains to be seen how the club will adapt to the rise of smartphones among its members, and what will be lost and gained through digital sociality at a distance.

Conclusion

The aim of this chapter was to offer an alternative to the critique of self-tracking as an individualised act comprising of a relationship between a user and a device,[52] by taking a broader look at self-tracking to reveal the importance of sociality, or warm contact, for self-care. This chapter has also sought to demonstrate that while mHealth might be considered a relatively new phenomenon, especially in its use among older populations, self-tracking and the gamification of health via visual markers have a long history in the Japanese context that precedes technological intervention.

This wider context of analogue health tracking through health checks, as well as the practice of keeping records and 'ranking up' on all sorts of activities, make the elderly Japanese population a prime target audience for mHealth. However, since my ethnographic evidence suggested a limited take up of mHealth apps, especially when compared to the ubiquitous use of the messaging platform LINE, the potential for generic applications for the purposes of health interventions must be explored.

Our small mHealth intervention demonstrated how commensality – the social practice of sharing food – can migrate online and in turn facilitate cultural knowledge transmission and the strengthening of social ties. The smartphone presents an opportunity to practise the culture of sharing (*osusowake*) digitally. As smartphones are connecting older people with their children and grandchildren who may live remotely, they are becoming increasingly important as vehicles for cultural sharing in a place where there is a fear that the wisdom held by older generations, especially pertaining to food and agriculture, is in danger of dying with them.

Taking inspiration from community health checks not as merely opportunities to produce biometric data, but as health practices that embed people in a web of social relations, the form of informal health-tracking evidenced in our intervention has the potential to re-socialise health and welfare in a context where there is great concern over the dissolution of community. In Chapter 7 I expand this focus on forms of community in Tosa-chō by looking at revitalisation efforts predominantly run by older adults which embed them in networks of care.

Notes

1. Lupton 2021.
2. NHK 2019, as cited in Danely 2022. In the article, elderly prisoners were reported to repeat minor crimes upon release as life inside prison was easier.
3. *Pachinko* parlours house slot machines and are widespread in Japan. They have existed since the 1920s as places for low-stakes gambling.
4. Tsutsui and Muramatsu 2005.
5. Campbell and Ikegami 2000; Tsutsui and Muramatsu 2005.
6. Suzuki 2015.
7. Aronsson 2022.
8. World Health Organization 2011.
9. Ajana 2017; Depper and Howe 2017; Fotopoulou and O'Riordan 2016; Lupton 2013a, 2013b; Mort et al. 2009; Rich and Miah 2014, 2017; Schüll 2016a, 2016b; as cited in Ruckenstein and Schüll 2017.
10. Lupton 2013a.
11. Kennedy et al. 2015; Lupton 2021, 187.
12. Ruckenstein and Schüll 2017.
13. Danely 2015b, 110.
14. Hänninen et al. 2021.
15. OECD stands for the Organization for Economic Co-operation and Development – a forum for policy making and collaboration for 37 democracies with market-based economies.
16. OECD 2024.
17. Statistics Bureau of Japan 2015.
18. Kimura et al. 2012.
19. Valtorta et al. 2016.
20. Cacioppo and Hawkley 2009; Shankar et al. 2011.
21. Wilson et al. 2007; Fukunaga et al. 2012.
22. Smith et al. 2018.
23. Tanaka and Johnson 2021.
24. Murakami et al. 2009.

25. Data of Tosa-chō 2021. http://www.town.tosa.kochi.jp/files/lib/3/353/202207111444356 087.pdf.
26. According to research by the Tosa-chō Social Welfare Council (*Shakyō* or *Shakai-Fukushi-Kyogikai*), Yamakubi 2019.
27. Ozaki 2014.
28. Gagné 2018.
29. Gagné 2018.
30. Dore 1973a, cited in Gagné 2018.
31. Anthropologist Iza Kavedžija has shown how such a community salon in Osaka can become a key site for the negotiation of interdependence in later life (Kavedžija 2019). See also Saito et al. 2019.
32. Nozawa 2015.
33. OECD 2019a.
34. OECD 2019a.
35. Kitayama et al. 2010.
36. Matsubayashi and Okumiya 2012.
37. Field Medicine started in Kahoku-chō, a neighbouring town of Tosa-chō, in the late 1990s by Prof. Matsubayashi. The project in Tosa-chō was initiated by Dr Yano Shoki from Tosa-chō, Dr Matsubayashi from Kyoto University and Dr Otsuka from Tokyo Women's Medical University (Matsubayashi and Okumiya 2012). Dr Kimura has been part of the organisational team on the health check for the past 11 years. The health check involves the collaboration of the mayor of Tosa-chō and the support from Tosa-chō's Social Welfare Office.
38. Kimura et al. 2008.
39. https://stats-japan.com/t/kiji/10714. Accessed 3 February 2022.
40. Nomura et al. 2017.
41. Ogasawara et al. 2013.
42. Daniels 2024.
43. Daniels 2001, 63.
44. Yanagita 1967.
45. The smartphone club was run by NTT Docomo, one of the main mobile phone operators in Japan, in partnership with the local Social Welfare Office. National commercial companies in Japan, such as the Japanese National Railways, often blur the line between private entities and public services. For example, the Nippon Telegraph and Telephone Corporation, which owns Docomo, was privatised in the 1980s, yet it is written into the company law that the Japanese government must own at least one-third of the total shares in the company (The Government of Japan, 2005). Thus the agendas and activities of mobile phone companies, such as the offering of smartphone classes for older adults, can be understood as closely tied with government agendas to increase digitalisation within everyday life, especially among non-adopters. Indeed, one of the key agendas of Prime Minister Suga, who came into office in 2020, was to increase smartphone adoption by reducing data tariffs to make them more consistent with global markets (Mingas 2020).
46. Haapio-Kirk et al. 2024.
47. Sasaki et al. 2021.
48. Czaja et al. 2017; Fang et al. 2018; Francis 2018.
49. Chopik 2016.
50. Czaja et al. 2017.
51. Grady et al. 2018.
52. Ruckenstein and Schüll 2017.

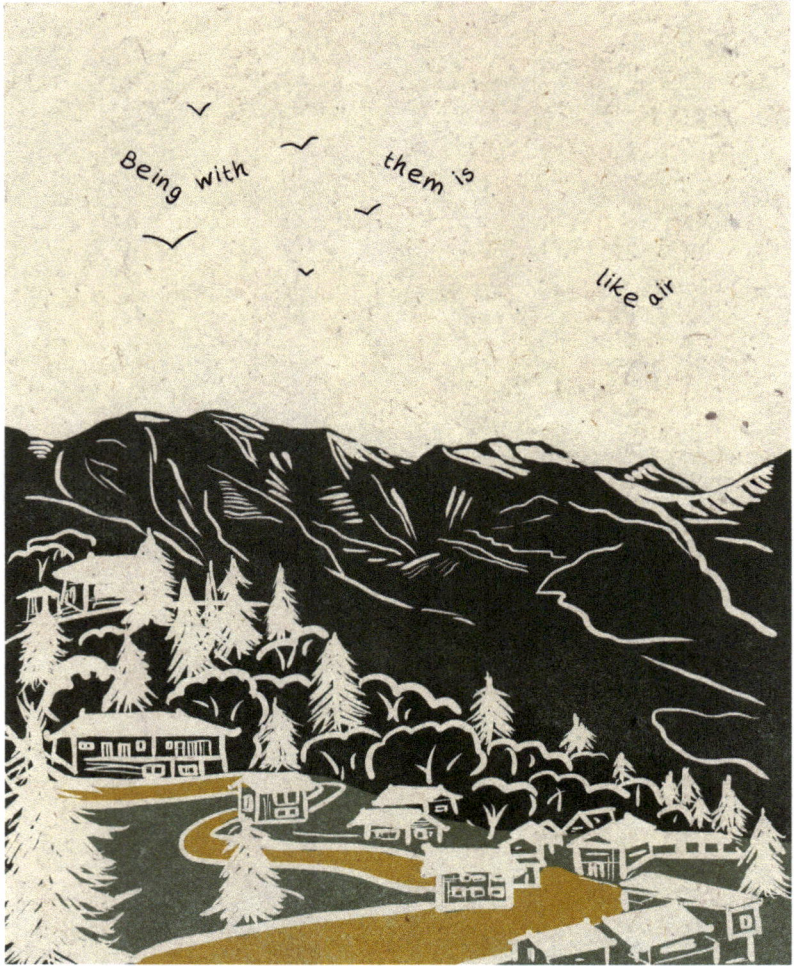

Image by Laura Haapio-Kirk.

7
Rural Japan: between decline and rejuvenation

Here gossip travels faster than the internet.

Miwa-san, aged 50, Tosa-chō.

I was standing at the top of a mountain about a 20-minute drive from the centre of Tosa-chō with Misako-san, an urban-to-rural migrant in her forties. The view down the valley stretched before us, revealing tumbling golden rice terraces ready for harvest, shining in the October sunshine. Suddenly Misako-san cupped her hands to her mouth and shouted 'ya-ho!' at the top of her voice, encouraging me to do the same. This was a special spot where you could hear your call reverberating off the mountain sides. Such echo spots, or 'ya-ho points', offer a rare opportunity for uninhibited noise-making in a culture where this is usually frowned upon.[1] Locals told me that before phones there was a tradition of shouting across mountain tops to communicate messages to neighbouring villages.

Recently high-speed internet access had been established across the region as a local government priority, partly in an effort to encourage migrants (*ijū-sha*) to the area. It was now more reliable to communicate digitally via LINE rather than through patchy phone reception in the mountains; there was certainly no shouting from mountaintops any more, except for fun. Small village communities that had previously been relatively isolated, resulting in fierce local pride and rivalry, were now connected via digital and physical infrastructures, such as roads tunnelled through mountains, that brought them into close and frequent contact. For locals and migrants alike, the radical compression of social proximity facilitated by digital technology has increased concerns about matters of closeness, distance, visibility and invisibility, all set within a culture of 'high-speed gossip', as put by Miwa-san at the beginning of this chapter. Other ethnographies in Japan have shown how this concern is

not limited to rural locations. It is equally applicable to densely populated urban sites, where people are concerned not to return to the over-
close relationships of the past.[2]

Rural Japan was sometimes romanticised as a locus of community by migrants to Tosa-chō, motivating their move from urban centres
where they felt less connected to people. Yet often the reality of a small
community's scrutiny quickly became experienced as intrusive, as discussed in Chapter 4. Migrants are thus situated between invisibility on
the one hand, due to their lack of local social capital, and hyper-visibility
as conspicuous 'others' on the other. Their behaviour and attitudes can
be provocative to locals, often diverging from traditional attitudes, for
example in relation to farming practices. When considered in relation to
the Japanese social concepts of inside/outside (*uchi/soto*) that have been
argued by some anthropologists to shape Japanese social interactions,[3]
urban-to-rural migrants are outside of normative ideals of what constitutes rural dwellers. However, they also represent one of the primary
ways in which super-ageing rural communities, such as Tosa-chō, can
survive into the future. Migrants are thus central to national programmes
of rural rejuvenation of depopulated areas (*kaso chiiki*) which this chapter addresses.

The Japanese social concepts of *honne* (true voice) and *tatemae*
(public facade) are useful when discussing how people present different versions of themselves according to the context (see Chapter 4). One
migrant in her late forties explained that it is easier to express her real
feelings through private messaging on LINE than on the more public
Facebook, where she was connected to people in the local community.
When she did post an update on Facebook that expressed her feelings,
she would suddenly feel embarrassed and worried that she had maybe
shown 'too much *honne*' for that platform. She was similarly careful to
maintain friendly but not too-close relationships in offline interactions
with other residents in her small village, half an hour's drive from the
main centre of Tosa-chō. As anthropologist Iza Kavedžija writes of downtown Osaka residents, people pay attention to politeness with each other
as a form of self-preservation:

> In a context where people are not seen as impermeable units, as
> discrete individuals coming together, but rather as constituted
> through social interactions, the distance created by attending to
> form is actually seen as vital for preserving the self.[4]

Migration[5] was seen as a different kind of self-preservation by migrants
who were motivated to seek community in rural areas because of the

precarity they felt in urban areas. This movement is a small countermotion to the massive and rapid out-migration of working-age people that has been occurring in rural Japan since the mid-twentieth century, some of whom are choosing to return in retirement. Such movement of people has led scholars to argue that 'Japan's rural regions have increasingly ceased to be *lifecourse* spaces ... and are instead *lifephase* spaces'.[6] By the 1960s and 1970s rural depopulation had become a major concern for the national and local governments across Japan. The shrinkage and decline of rural communities was the consequence of various concurrent processes, at the forefront of which was the loss of working-age people to cities, combined with demographic ageing.[7] As discussed in the focus on health in Chapter 6, rural depopulation prompts concern for those left behind. Population decline and ageing are not limited to rural areas, even if it is in the countryside where they are most acute. By the turn of the millennium, widening inequalities associated with both demographic trends were regarded as characteristic of contemporary Japan as a whole.

This chapter explores how population ageing and shrinkage have impacted the lives of local people and those who have chosen to migrate to Tosa-chō as part of a current wave of 'lifestyle migration'[8] occurring in rural areas across Japan.[9] Through ethnographic examples of community and individual efforts at rejuvenation, I show how civic action situates people within intergenerational social networks of care. I critique the idea of the rural as a 'life stage' space in which people are born and raised, and then perhaps retire.[10] While many older people I met in Tosa-chō retired back to the region after working in cities (so-called 'U-turners') to spend their later years in their hometown, for younger urban-to-rural migrants (or 'I-turners') the rural setting provides opportunities for a self-sustainable lifestyle that they feel is not possible in cities. For some this move is temporary. Others intend to remain in the countryside precisely because they can envisage it as a space in which they find belonging and meaning.

In this chapter I show how digital technologies are emerging as key sites in which migrants and locals alike calibrate their sociality with each other in a rapidly changing social context. In illustrating how older local populations interact with the younger migrants once they have arrived, I argue that the older resident generations find both fulfilment and meaning from interacting with and helping migrants, often of a similar age to their own children or grandchildren, many of whom have left for life in the city. In turn, the migrants play an important role within normative intergenerational practices that are significantly challenged by depopulation and an ageing society.

Rural rejuvenation

Rural rejuvenation has become a key priority of local governments in rural areas across Japan, and indeed of policy at the national level, to tackle the processes of marginalisation that comes with being at the forefront of a 'super-ageing' society.[11] In rural studies research has championed rural communities as pioneers in dealing with demographic changes, both in Japan[12] and in other ageing countries such as the UK.[13] The term *genkai shuraku*, meaning 'marginal village', was coined by Japanese sociologist Ohno to describe communities in which 50 per cent of the population is aged over 65.[14]

While Tosa-chō is not quite at the severe status of *genkai shuraku*, a process of community revitalisation is currently underway to maintain the population and avoid a slide into 'marginal' status. Revitalisation efforts of similar communities in nearby prefectures have included the creation of new business ventures, such as employing older adults in decorative leaf production for lunch boxes (*obento*) in neighbouring Tokushima Prefecture[15] or tourism-based initiatives in the nearby Seto inland sea area.[16] In Tosa-chō a mix of revitalisation approaches include the promotion of tourism with the refurbishment of a large hotel and the creation of holiday cottages, the creation of business ventures involving older people, such as tea leaf picking, and a concerted effort by the town hall to recruit and support migrants. These efforts involve both centralised ventures from the local government and more local initiatives by village chiefs and volunteers. This chapter will leverage the advantages of an ethnographic study by focusing on the consequences of such developments for residents. It will explore the experiences and motivations of individual volunteers involved in rejuvenation efforts, typically in their sixties and seventies, and consider the relationships they develop with younger migrants.

Many 'I-turner' migrants were being, or had previously been, supported by the Regional Revitalisation Co-operation Officer Program *(chiiki okoshi kyōrokutai)*. This is a nationwide initiative that financially supports migrants seeking to relocate to rural areas and start business projects.[17] The activities covered by the scheme are wide-ranging; I met people who had received support to open restaurants, practise farming and run tourism initiatives. Some were even being funded to learn local crafts such as woodworking and paper making. The scheme supports people for three years; due to the low cost of living in rural places such as Tosa-chō, entire families could reasonably live on the earnings of these projects. The Program allowed migrants to start their businesses as soon as they settled in the area, rather than having to spend time looking for work. Because of the extended nature of the funding, it gave migrants time to establish relationships and networks in the community before their funding ran out.

For some participants the scheme proved highly successful. They were able to take on ambitious projects, such as renovating an abandoned school to become a community centre and hostel, which then continued as a viable project beyond the funding period. Others criticised the scheme, however, declaring that it did not make sense for the government to fund projects that would otherwise be unsustainable. They argued that for these migrants to stay in their chosen rural locations, their businesses needed to be self-sufficient once established. Some people worried that the funding encouraged migrants to drift through such positions, moving from place to place on short-term projects that never resulted in the long-term goal of permanent relocation, the real purpose of the scheme.

Indeed Kawamura-san, a local man in his seventies who was involved in community planning, explained that about half the number of migrants who come to Tosa-chō on this scheme left after their three-year period was over. For migrants to settle more permanently, he explained, there needed to be more sustained support for businesses, including those not owned by migrants, to encourage local young people to take over family businesses rather than to leave for the city in search of employment. In the same way that farmers receive subsidies, he believed that more long-term financial support was necessary to keep young people in this rural town. You can hear Kawamura-san talk about these issues in the video that introduces the fieldsite in Chapter 1 (Fig 1.4).

While many rural areas are experiencing decline, the Japanese countryside has retained a dynamism that challenges the notion of the 'peripheral' rural in binary opposition to urban centres.[18] Rural communities are places through which new ideas, practices and people flow.[19] In light of the Covid-19 pandemic and the associated surge of interest in the kind of alternative lifestyle that the countryside can offer,[20] rural populations are at the forefront of innovation regarding wellbeing, care and intergenerational sociality.

Kōchi Prefecture, where my rural fieldsite is located, can be considered one of the most peripheral in Japan, given its remote geographic location and the long-term economic and demographic decline associated with the region.[21] The towns of Tosa-chō and neighbouring Motoyama-chō are in the moutainous north of the prefecture, situated along the Yoshino river; they are connected by a main road where the amenities such as schools, hospitals and supermarkets are situated. Driving through these central areas, one does not feel the reality of depopulation that is facing this region, with the local population having halved since 1965. It is only after venturing off the main road and into the mountains that one finds whole villages and hamlets abandoned, along with empty school buildings and rice fields that have not been tended for many years. In these places the remaining population, mostly older individuals, worry what will happen if they no longer have neighbours to greet and cannot rely on the reassurance of having people living nearby.

Locals in their seventies and older have memories of walking for hours between villages to attend school or buy coal. Today, however, everyone relies on cars to reach public facilities as there is limited public transport. For older people who lose the ability to drive, there is real potential for isolation. When rural communities had larger and more sustainable populations, as during the early-to-mid twentieth century, even remote towns such as Tosa-chō had multiple schools and facilities. Following the decline of forestry and agriculture across Japan, many people of working age left rural areas for employment in cities rather than continue their parents' agricultural businesses.[22]

As one local man in his seventies explained, 'there are more and more houses without lights on'. The concept of seeing houses with lights on appeared as a positive motif; it was expressed by different people and associated with feelings of security and protection. Anthropological research on home lighting in Japan reveals bright lights to be linked to feelings of safety and modernity.[23] A migrant from Kanagawa who lived with her young family in a house surrounded by rice fields said that she felt comforted at night by the lights of a neighbouring couple in their eighties who lived nearby. Through spending time together, and learning skills from the older couple such as how to dry persimmons, they had become like parents to her and grandparents to her children. Both generations appreciated the comfort of friendship and local support. People fear that having few neighbours will make life harder for elderly residents because the community had traditionally been sustained by *yui*, a concept of mutual forms of care and assistance discussed in Chapter 4.

Yamakubi-san, the head of Tosa-chō's Social Welfare Office, explained that facilitating and maintaining supportive informal infrastructures of care, described in more detail in Chapter 6, was one of their top priorities. Maintaining a supportive community as a safety net was seen as essential in helping people to live independently at home into old age. In the early 2000s, when the national government moved away from a system of central support for rural areas suffering from declining numbers of taxpayers, depopulated regions were forced to become more self-reliant.[24] This was the era after the burst of the bubble economy in the 1990s; it heralded a period in which funding for local services in rural areas significantly diminished. Yet in Tosa-chō it was precisely this sense of urgency around decline that galvanised people into community action. This in turn provided a source of activity that became central to the lives of older volunteers, as we shall see below.

Missing generations

Many middle-aged children of Tosa-chō's elderly residents had moved to cities for work. It was therefore common for older people to live alone or with their partners, while their children and grandchildren lived several hours away. With this dispersal of families, there was a widespread feeling that there were few opportunities for older people to pass on their experience and skills to the next generation, for example knowledge embedded in the daily practices of farming and caring for the land. Such a belief perhaps explained the enthusiasm of older people for welcoming younger migrants to the area and for becoming involved in activities that would bring them into contact. Conversely, I heard countless stories that confirmed the eagerness of migrants to learn from older local residents. These often resulted in familial-like relationships with older mentors as Megumi-san, a migrant from Tokyo in her forties, explained:

> Now the elderly are able to teach them [the migrants] so many things about cooking, farming you know? So the community is coming to life again. That role of teaching and learning – in Japan, in the past that's how society functioned everywhere, from home to home. But that culture is dying. The culture of teaching from generation to generation. With the high economic growth that culture diminished. So the immigrants have this important role in this town. To listen and learn, which uplifts the locals.

Megumi-san's three-year old daughter was growing up far from her own grandparents in Tokyo, yet she interacted regularly with older local people and spoke with a local Kōchi dialect. Megumi-san posted photos of her daughter foraging mountain vegetables (*sansai*) with a local older man, along with a comment that he had become like a grandfather to her. Many migrants had close relationships with locals, whom they would describe as their Kōchi family and adopted parents or grandparents. Often these relationships were centred on learning traditional practices, such as how to preserve fruits and vegetables, or on foraging in the mountains.

Not only do older local people take on the role of mentor, and indeed express forms of fictive kinship, by letting migrants accompany them on their own foraging or farming activities; some also reconfigure their land to accommodate migrants who want to practise small-scale agriculture for themselves. One local man, who had retired back to Tosa-chō after working in Tokyo, had dedicated several plots of his land for use by migrants. He rented this land for a small fee and provided advice and support to the migrants, who often had little experience of agriculture. In a shed he had pinned up detailed instructions and seasonal advice about when to sow and harvest a wide variety of different vegetables, such as Japanese radish (*daikon*), potato and pumpkin (Fig. 7.1). The information included photos of each vegetable and a translation of the vegetable's name into English, due to him renting land to several international migrants.

While this man also practises farming alongside the plots that he rented out, some locals expressed little interest in the hard labour of cultivating their fields themselves. They were much happier at the thought of renting them out in their entirety to migrants, together with the promise of sharing some of the crops. I participated in one such harvest of a

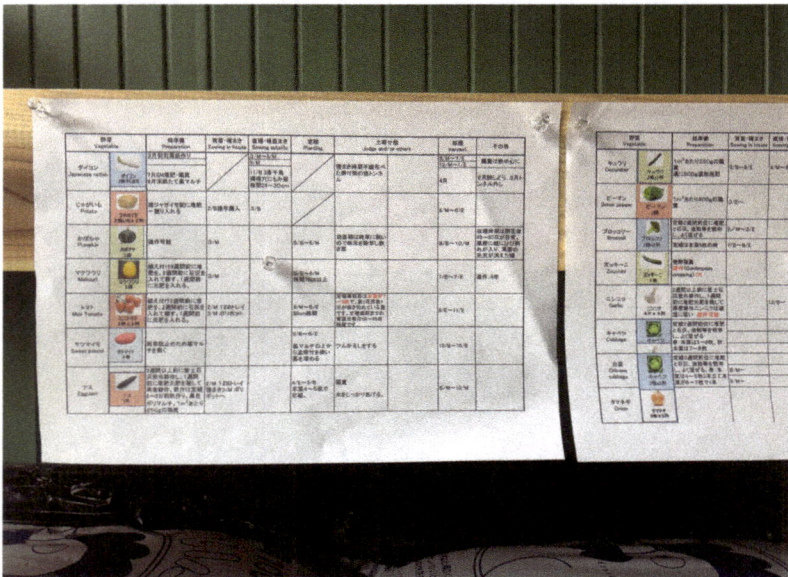

Figure 7.1 A photo of instructions for farming vegetables. Image by Laura Haapio-Kirk.

local family's potato field. The work was long, intense and muddy. After loading crates on to a van the harvesters – all migrants, and in one case a potential migrant, who was making up her mind whether to move to the area with her young son after a divorce – reconvened at the family's house to share lunch and rest. Before leaving, everyone filled sacks with potatoes (enough for several months). The family also allowed the harvesters to pick from their crops of pak choi, daikon and cabbage. Participating in local agriculture and knowing the source of their food was one of the key reasons why many migrants had chosen to leave their urban homes and move to the countryside. This is a wider phenomenon now occurring across Japan and it will be discussed below.

Forestry was historically one of the main sources of income in Tosa-chō. Now, however, the trade is in decline due to the availability of cheaper imported wood from China, contributing to the depopulation of the area. Despite this decline, a connection to the forest and mountains remains strong for many people, especially those living in more remote areas. For example Midori-san, a man in his seventies who practised farming and woodworking, lived roughly an hour's drive away from the centre of town. He spoke to me about the importance of care and community in the development of the individual, significantly drawing upon the lives of the trees he knew so well:

> My grandchild is in the last year of high school now and they have to choose the path of their life. The trees are same. When you are in your early twenties, you have to decide what kind of person you want to become in society. Trees too. A tree that is 18 will tell us if it's going to eventually become a house or if it's going to bend, like that. And who supports that development? The parents. The owners decide what's best for the tree. When the tree is in its thirties to forties, it will not survive in the harsh world in the mountains if it has not been supported. Same with humans: if your parents don't teach you right in the first 20 years of your life, how will you survive living in society? In your twenties, trees and humans are both flexible, but when you're in your thirties or forties it's harder to change your ideas and adapt.

Midori-san, who appears in the film in Chapter 8 (Fig. 8.1), based his ideas about life stages on an understanding of age-based responsibility. According to him, it is the duty of older people to show younger generations how to behave in society and how to become a productive individual. However, in rural Japan young and middle-aged people are rapidly leaving

villages to go to work and live in cities, so depriving the elderly of opportunities to communicate or socialise with younger people; they are thus
unable to fulfil their roles as senior mentors. Midori-san lived in a very
remote spot. A town hall worker, worried that my research assistant and
I would not be able to find it, printed off and Sellotaped together five maps,
showing the complicated directions to his house (Fig. 7.2). Following the
hand-annotated landmarks and directions, we were taken deeper into the
mountains, passing occasional single houses, some abandoned, others still
inhabited. A woman, perhaps in her eighties and bending over her plants in
front of her house, stopped to watch as we drove by. In places such as this
there were few opportunities to meet other people, let alone children.

The quote above illustrates the importance of multigenerational
sociality for shaping the beginning of a life – yet it is clearly also vital
towards the end of one. Often having little direct experience of spending prolonged time with their grandchildren, it can be difficult for older
adults to know how to be around young people. As Midori-san explained
at the lunchtime event introduced above:

> Mums are encouraged to come with their kids to this lunch event
> and the elderly can come to hang out too, but they don't know how

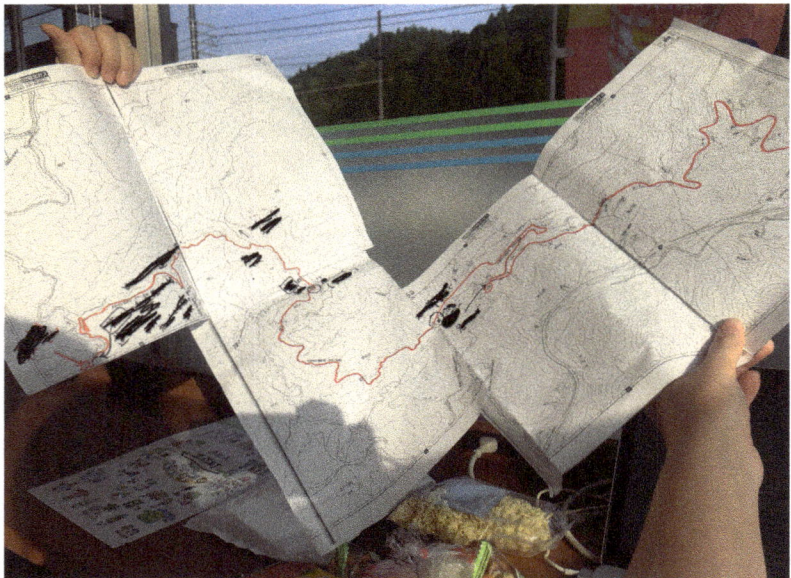

Figure 7.2 A hand-annotated map of a remote region. Image by Laura
Haapio-Kirk.

to! It's not like they took care of their grandchildren either, you know? It's hard. There is a generation gap so the elderly haven't had the chance to teach their skills to anyone else.

The concept of *yui* relies upon contact between individuals who can then form bonds of mutual support, something that such events are designed to encourage. Restoring the flow of knowledge and reciprocity was seen by many migrants and older locals alike as a way of saving the local culture, threatened by the demographic decline facing the community. While this rural town could itself be understood to be in a state of precarity because of depopulation, migrants were often moving from what they saw as even more precarious lifestyles in cities. In so doing, they were often motivated by precisely the form of community support that they might build with locals.

Precarity and uncertainty

In Tosa-chō many migrants explained how they had been motivated to move from cities and seek a rural life for themselves and their families after the earthquake, tsunami and nuclear meltdown of 11 March 2011 (known as 3.11). Chiho-san is originally from Tokyo and was there when the earthquake hit. In that instant she picked up her children, one under each arm, and made the decision then and there to relocate, seeking to protect them from what she felt was an insecure life in the city. Chiho-san felt overly reliant on what seemed like unstable infrastructure in the city that she had no control over, such as her food and water supply, making her long for a more self-sufficient lifestyle.

Others made this same decision in the aftermath of the disaster, when the fear of radiation contaminating food and water became overwhelming. City life represents convenience, epitomised by 24-hour supermarkets, but for precisely the same reason the city has made people over-habituated towards convenience, experienced by some as a loss of autonomy. By relocating to rural areas, people have an opportunity to grow their own crops and to feel more connected to the source of their food, even if they are not directly producing it. Anthropologist Susanne Klien argues that the breakdown of trust in the state, concerns surrounding radiation and food safety and the period of national reflection that followed 3.11 accelerated an urban to rural migration that was already happening with increased urgency following the 2008 financial crisis.[25]

Migrants to Tosa-chō were attracted by the ideals of community that the region offered, in contrast to the fracturing communities that they had experienced in the city (discussed in Chapter 4). In the economically uncertain context of contemporary Japan, they were also drawn by the low cost of living in rural Kōchi Prefecture. Chiho-san explained that one of the things that she loved about her new home was that even if you did not have money, people in Kōchi were rich because of the abundance of the land and the ethos of sharing produce that dominated this community's social life. Another 'I-turn' migrant, Kawase-san, aged in his forties, explained that 'another kind of economy is running in this town' – one that relieved migrants of the incessant need to pursue well-paying jobs in the city, enabling an alternative existence to take shape in the countryside. I came to experience this sharing directly, often participating in the circulation of goods as people generously gave me far more than I could reasonably bring back home to Kyoto. I was in regular receipt of produce that people had grown and harvested themselves, such as rice, *yuzu* citrus fruit and chestnuts.

As discussed in Chapter 4, the concept of '*yui*' (mutual support) was framed as stemming from an agricultural past before machines, when people relied on labour from their community. One five-generation family who lived in a more remote and mountainous part of Tosa-chō would gather their extended relatives and friends for sowing seeds and harvesting crops at various points throughout the year. Many people joined these activities so that they became known as mini-festivals. It also became a tradition for the participants then to move on to the next family to help with their harvest, so that everyone's crops in the area received help. For this family and many others, an informal economy of care in the form of voluntary farming labour in exchange for return labour and the sharing of harvested crops was in operation. While machines can do the work of many people, these activities persist because of the benefits for community life.

Rather than simply being a form of nostalgia for pre-mechanised farming, however, a wider sense of nostalgia for the countryside[26] can be understood as emerging from the historical migration patterns within Japan. Kiomi-san is in her late thirties and an academic in Osaka. Her family story is typical of many families in Japan, with many baby boomers, such as her parents, having moved from the countryside to cities to work for big companies during the 1970s and 1980s. This movement of people out of rural towns to urban areas became extremely rapid during the last half of the twentieth century.[27] Kiomi-san explained:

> If you come to my home, my family is so typically Japanese. You can see the industry shift. My grandfather was a farmer. My father left

home, went to university and worked for a semiconductor manu-
facturer. I went to do a PhD. It's very typical! You can see other families like mine in many places. In the 1980s, my father's generation,
the industry changed dramatically. We feel nostalgic (for the countryside) because it's the life of our grandfathers and grandmothers.

For the 'I-turner' migrants in Tosa-chō, this nostalgia for the idealised
community experienced by their grandparents' generation is in direct
contrast to the way that the city makes now them feel. The superficial
veneer of comfort and convenience offered by the city was pulled from
under them on 3.11, revealing a sense of deep vulnerability and isolation.

While 3.11 served as a catalyst for migration, the feeling of a lack
of connection to their urban homes was of longer duration. Hiro-san,
a migrant from Tokyo in his forties, explained that cities such as Tokyo
were not healthy because they were too 'far from nature', although his
conception of nature was entirely centred on long-standing human habitation and a notion of belonging linked to ancestry. Together we visited
an abandoned hamlet about a 30 minute drive away from the centre of
Tosa-chō. As we walked past houses that had wood shuttering the windows, I noticed how the flowers planted by previous inhabitants were still
blooming. Stopping to talk beside a stone wall, Hiro-san explained the
difference between living in a city and living in a rural area:

> When people live in a place where it's so far from nature, then they
> think life is made by themselves. But a place like here, life is made by
> nature or by generations long ago. You see a stone wall here, maybe
> your grandmother might know who made this wall, you know?
> Tosa-chō people do feel they are connected to many generations. If
> you grow up in the city, you don't feel like that. When your previous
> generation made your life and you can really feel that, then you are
> conscious that your children will carry on the next generation. So
> you're more forward looking. What can the next generation do, you
> know? What kind of town can we leave for the next generation?
> What can I do for the town? You feel responsibility.

It was often a sense of belonging that migrants were searching for. This
was something that they felt to be lacking in cities, which change so rapidly and might have often only been inhabited by the last couple of generations of a family.

Belonging, according to anthropologist Michael Jackson, is in fact
an existential concern related to finding one's place in the world[28] and the
ability actively to shape it.[29] For Hiro-san, without a historical connection

to the area, belonging emerges from working to protect his environment. In the city, where things change so fast he has less of a sense of the human story behind his surroundings and feels alienated. It was poignant that he told me this as we stood in a hamlet that no longer had any residents at all. Living in Tosa-chō, somewhere where he experiences belonging, makes him feel a sense of responsibility towards his new home, along with a determination not to let it slip into demise. Hiro-san now works with the local town hall (*yakuba*)running various community rejuvenation projects. His new life gives him a sense of belonging and purpose[30] that he lacked in the city.

'I-turners' are also pushed to migrate by the stress and overwork that they experience in cities, which require high salaries to meet living costs. Hiro-san explained:

> In Tokyo, it was so hard to distance myself from money. I was chased by how much I was earning, (and having to worry about) paying rent. Here I am more relaxed – it's nicer. You don't need money and everything is so cheap here. And food too, there are many people who feed you! Ha ha ha! But there aren't many jobs, and even if you get a job the pay is so low. But you can still live here because everything is cheap. I am definitely much happier here.

After the bubble economy burst in the 1990s, workers could no longer rely on lifetime employment and the often-maligned phenomena of part-time and casual work (the so-called '*furita*' lifestyle) emerged. The decision to move to rural areas can allow people to break free of the pressures of seeking and maintaining a long-term company job, especially for men, enabling family life finally to become a priority. The lifestyle they envisage as being possible in the countryside is in opposition to the lives of their parents' generation, for whom lifelong company loyalty meant little time for their families.

Masa-san was in his mid-thirties and had moved to Tosa-chō with his young family from Osaka. He worked for the same national company that he had worked for in the city, which had a branch in Tosa-chō. Every day he would endeavour to come home on time and spend the evenings with his daughter and wife. In Osaka he felt pressure to work long hours and then go out with his colleagues after work. He remembers his own experience growing up with a father who was always working. He did not want to be like that with his own child:

> He looked depressed and really exhausted and we couldn't talk to each other about his day off. When I was a kid, I used to ask my father

'let's go somewhere' or something like that, but he would say no, he's too tired. That I should go by myself. And I don't want to be like that.

As mentioned in Chapter 3, the new 'type' of man that Masa-san represents, known as *ikumen* and who actively participate in childrearing,[31] are emerging in opposition to the traditional patriarchal structures that previously dictated family life. Satomi-san is in her mid-thirties and has a three-year-old daughter with Masa-san. Satomi-san explained that if her husband was not like this, an *'ikumen'*, then she would not have considered having a baby with him. She suggested the solution for Japan's low birth rate is for more men to embrace the *'ikumen'* approach and find a balance in life between work and family. I suggest that the countryside can be a place where such new identities can flourish, while of course recognising that such migration does not completely remove people from oppressive social structures they hoped to leave behind in the city, nor indeed from responsibilities of care.

Lifestyle migration

The Sawada family moved to Tosa-chō from Tokyo to raise their young family in what they consider to be a traditional and healthy way. They shared many aspirational values with earlier waves of 'lifestyle migrants' who sought to practise farming and live different lives to their salary-driven lifestyles in cities.[32] 'Lifestyle migration', as defined by sociologists Michaela Benson and Karen O'Reilly, frames this form of spatial mobility as involving relatively affluent individuals seeking a better quality of life, community and lower costs of living in places that are meaningful.[33]

The Sawadas bought and renovated an old wooden house in a remote area and installed a composting outdoor toilet and a wood-fired bath. Their kitchen has an original wood-fired stove where they prepare meals with locally grown produce. Their kitchen tap only runs cold and they do not use washing-up liquid, meaning that oily dishes have to be scrubbed with sand, a system that results in ice-cold hands. Their minibus-style car, used to transport their five children, runs on recycled cooking oil; it is easy to tell if the family have just driven by because of the smell of fried vegetables that follows the vehicle. This family is passionate about permaculture (a self-sufficient and sustainable approach to agriculture) and do not believe in using pesticides or chemical fertilisers on their crops. The clothes worn by the Sawadas' children are all handed down to younger siblings, and they try to spend as little money as possible. They earn money through having occasional visitors who come to stay at their home to see whether the 'permaculture lifestyle' is for them. In every way imaginable this family were striving for a minimalist, low-impact lifestyle that would enable them to live and eat well, and their children to grow up as far removed from city life as possible.

To fulfil this dedication to living life in a way that would not be that different to a century ago, the family relied on smartphones and an iPad on a daily basis. They used social media including Facebook to build a network of people interested in converting to this lifestyle, advertising their home as somewhere that people could come and visit. I stayed with them during the visit of one such family. A young couple with a two-year old son had travelled from Hiroshima to experience a night in the Sawada family's home. We ate dinner together, all prepared with organic local produce, and the visiting couple asked questions about the family's life there. Afterwards we played card games with the children and then took turns having a bath in their wood-fired tub, before rolling out our futons and trying to sleep, accompanied by the continuous calls of frogs in the nearby field. The next morning I asked the visiting father why they were considering leaving the city for the countryside, to which he simply replied 'City life is not a happy life', as if he was stating something very obvious.

In anthropologist Susanne Klien's recent research with migrants aged between 20 and 40, who had moved from urban areas to rural Tōhoku in north-east Japan,[34] she found that they were motivated by the same set of ideas and contemporary phenomena: diversifying lifestyles, self-realisation, living for oneself, a better work/life balance and living for the moment rather than striving for future stability. Such motivation is not new. Even during the 'bubble economy' years of the 1980s there were urban to rural migrants known as 'salary rejectionists' (*datsusara*); members of the 'back-to-the-land movement', they moved to depopulated villages in the Kii peninsula.[35] So-called 'lifestyle mobilities',[36] as opposed to other forms of migration such as that driven by purely economic factors, is often defined in the academic literature as being motivated by choice rather than necessity.[37] However, the migrants I met in Tosa-chō had taken the decision to migrate because a mixture of both choice and necessity. The precarity they sensed in urban areas pushed them to make drastic lifestyle changes towards self-sufficiency, changes that they felt were necessary for the wellbeing of themselves and their families.

The Sawada family seemed very settled. However, one of the problems of the 'I-turn' (urban-to-rural) phenomenon is that this population, who are themselves approaching middle-age, then face the problem of their parents becoming old and dependent far away in a city. As Sawada Kenji-san explained:

I don't know if we will stay here forever. Now it is great and my parents are healthy, but who knows in the future? And who knows when we get old and need medical care too – maybe we would move back to the city. I don't know.

There are rumours that the town will build a care home for the parents of migrants who are moving there. However, the real issue was that older people often do not want to move to a new area at such a stage in life, leaving behind friends and familiarity in the city. One person confided that he thought this new care home was just a plan to boost population numbers since he believed the town hall receive a financial reward for every person who migrates to the town. Such criticism reflects a persistent belief that government schemes are not necessarily about what is best for the individuals involved, but rather about achieving targets and following state agendas.

The 'lifestyle migration' exhibited by the Sawada family is defined as migration within the developed world among affluent people who relocate to seek a better way of life.[38] In Tosa-chō migrants often began their new rural lives with greater economic capital and broader social networks than local people, connecting them to centres of power in Japan and internationally. It was thus not surprising that some younger local people expressed resentment that migrants were being supported by the rejuvenation schemes to pursue projects that to them often looked more like play than work. One local woman in her forties worked at five jobs to support her family. For her, the lives of the migrants were completely removed from her daily reality. Other locals remarked that if migrants really wanted to 'live a rural life' they should take jobs in supermarkets and convenience stores to experience the kind of work that native locals do. Kōchi Prefecture is the second poorest in Japan. Rural areas face the brunt of financial hardship; there are few jobs, and those that are available tend to be low paying. Such is the reality behind a romanticised and nostalgic notion of rural life that fuels many of the decisions to migrate from the city. Often these visions are far removed from the daily life of the people living in these areas, and migrants soon become aware of these tensions.

Physical and digital migration

While many 'I-turner' migrants evaluated their move to rural Kōchi Prefecture as a positive decision because of the friendliness of local people and the warm relationships they had established with often older local members of the community, they simultaneously experienced the ostracism inherent in the Japanese experience of being an outsider, particularly from the younger local people who had never left the town. Middle-aged local people spoke of struggling to accept the newcomers; they viewed them as perpetual outsiders because of their lack of family ties and unfamiliar styles of dress and behaviour. One local woman in her fifties described how she did not even see the migrants when in the supermarket; their lack of shared social and historical ties rendered them invisible despite, or perhaps because of, their conspicuousness. In this context of social invisibility, smartphones become an important way for migrants to connect with one another and for migrants and locals alike to manage their visibility and find belonging.

Many people in both my urban and rural research strongly felt that their smartphone was their private territory, much like the physical home. As the extensive work on Japanese domestic space by anthropologist Inge Daniels demonstrates, certain features of the traditional Japanese home, such as enclosing fences and high walls, frosted glass and barred windows, are designed to keep the interior private space separate from the public exterior.[39] However, as Daniels argues, these divisions are not static but in a constant state of negotiation. The smartphone similarly represents the most private and most public of spaces all at once, while the borders between the two are constantly negotiated and in flux. I attended workshops in the homes of migrants that took on a semi-public ambience; we learned about tea blending and practising calligraphy, while also enjoying home cooking and the relaxed atmosphere of a home setting. The opening up of migrants' homes to other migrants as semi-public spaces was central to the sociality of the migrant community – perhaps because there were few places in the town where they could gather. Yet by conducting workshops and get-togethers in homes, these events were less accessible to those outside the migrant community. The material environment contributed to the divide between insider and outsider, which was then replicated online.

In the research it was common that people might feel comfortable with a certain degree of intimacy on one app which might then not be appropriate for a different app. I observed that Tosa-chō migrants and locals tended to congregate online in platform-specific ways. Migrants tended to participate in communities on Facebook, whereas local people more often used only LINE groups and private messaging. People's online interactions were thus conducted in ways that reflected the separation of the two communities offline. Migrants tended to have wider social networks that also extended internationally, motivating them to use Facebook. Maintaining personal boundaries and calibrating appropriate levels of social distance was a concern for migrants and locals alike. Itsushi-san, a migrant in his forties who had married into a local family and therefore had an insider/outsider position, was concerned one day when LINE synced with his phone book and invited all his contacts to connect with him on LINE:

> LINE scares me a lot, you know. Once or twice I made a mistake and made myself visible for many people through my phone book. And it sent an invitation to connect through LINE to many, many people and that really scared me! And then I heard back from a few, but you know how you can change your name on your LINE profile? Yeah, there were many people that I didn't even know who they were! It was really scary. It should ask before you share your information online.

Itsushi-san explained that he believed LINE puts you in danger of

> collapsing relationships of trust ... I think LINE does that sometimes. Facebook Messenger is okay because it doesn't connect with my phonebook ... LINE deepens the relationship with another person too easily and too quickly, you know? It deepens the relationship with someone you don't necessarily want a deep relationship with.

For Itsushi-san, Facebook messenger was better than LINE because he would first have to accept a friend request in order to chat with another person, avoiding any risk of deepening a relationship unexpectedly. Miller and Turkle have separately proposed the 'Goldilocks strategy'[40] or the 'Goldilocks effect',[41] in which people use affordances of social media platforms to keep people at just the right distance. However, Itsushi-san's experience with LINE demonstrates how technologies can sometimes subvert people's desires regarding levels of sociality, the effects of which can be amplified in a small rural community where people are already highly sensitive about their visibility.

I found that 'U-turner' migrants, often retirees returning to their hometowns after working in cities, were situated somewhere between local people who had never left and I-turner migrants in terms of the platforms they were on and how they used their digital connections to seek belonging in the homes to which they had returned. They preferred to build modern homes which would allow them to continue the comfortable lifestyles they had become accustomed to in the city. By contrast, I-turner migrants who desired 'authentic' rural life were involved in renovating traditional wooden homes and sought low-cost, often basically free, accommodation in abandoned homes. Takeshi-san, a man in his early seventies who had moved back to Tosa-chō upon retirement, was motivated by a desire to live closer to his brothers who had never left. His wife had passed away just before he retired, making him value his family connection more. 'Being with them is like air,' he told me, 'it is something you can't live without.' However, he also could not live without his digital connections.

'Without a computer I cannot live here. Because it's so isolated. I'm not lonely, but I'm using my computer for so many purposes. Buying, mail, editing photos, websites and such,' Takeshi-san explained. His house was high-tech. He had a Roomba robot vacuum cleaner and the toilet, bathtub and several kitchen appliances were all digitally programmable: 'many things in my house can communicate with me,' he said. He relied on his computer to Skype with his daughters who lived abroad and used Facebook on his smartphone for calling because it was cheaper than calling on his *garakei*, which he still maintained. He was more comfortable using his computer than his smartphone for most tasks, since he had only recently inherited the device from his late wife. In contrast he had been an expert computer user for his whole working life, even building his own computers.

For Takeshi-san, building a house that had lots of space for socialising was important, including a terrace with a barbecue and a large open living room/dining room/kitchen. Physical social proximity to his brothers and the communitiy he grew up in was Takeshi-san's main motivator for moving back to the region. Yet it was only through digital means that he was able to live there and feel connected, not only to his previous city lifestyle and to his family living abroad, but also with the people living right next to him. He would often post photos on Facebook of crops he had harvested from the plots next to his house. In one photo, of an abundance of tomatoes, he asked for recommendations of what to do with them. A local neighbour replied with a suggestion to make pasta sauce and Takeshi-san subsequently invited this neighbour over to taste the results of his efforts. Takeshi-san is a typical example: like many people, he recognises the importance of physical proximity for sociality, especially now that he is older, yet this physical proximity is often premised on digital connection.

Conclusion

This chapter has discussed how population ageing and shrinkage have impacted the lives of local people and migrants to Tosa-chō, complicating the picture of rural Japan as a 'life stage' space only associated with later life.[42] 'Lifestyle migrants'[43] in their thirties, forties and fifties are choosing to move to rural areas because of a desire to raise families and retire in communities and landscapes that offer a sense of security through both self-sustainability and mutual support. I have argued that older residents benefit from such migration because it can restore the possibility of inter-generational contact, often lacking after the residents' own children have moved to cities. For those middle-aged and younger locals who remain, however, their comparative lack of job opportunities and financial precarity can lead to resentment of migrants, who are often supported by state rejuvenation schemes and who benefit from greater social and financial capital.

Building on recent work by other anthropologists in Japan such as Klien[44] and Rosenberger,[45] I have demonstrated how rural towns, such as Tosa-chō, are situated on a rural-urban continuum: they both receive people who are disaffected by city life and lose others who are seduced by the possibility of better paying work in the city. Such flows of people have made Tosa-chō a highly dynamic place because of the deep sense of civic purpose and responsibility towards rejuvenation that many people share. Prompted precisely by the danger of peripheralisation associated with depopulation and population ageing, locals and migrants are actively engaged in manifesting belonging through rejuvenation efforts. I argue that rural decline affords possibilities for both civil action and personal fulfilment, providing a landscape in which people experiencing various forms of precarity and alienation may find community and belonging.

However, the picture is complicated by the 'conspicuous invisibility' that surrounds urban-to-rural migrants. The smartphone has become a space in which people manage their visibility, inflected in Japan with the social tendency of hiding one's true feelings behind a facade (*honne/tatemae*). Despite the challenges facing migrants and locals as they adapt to social shifts resulting from rural depopulation and efforts at rejuvenation, people seeking a sense of belonging in the countryside are aided by the convenience and connectivity that digital platforms provide. Rural migrants and older local people share feelings of responsibility towards their surroundings; they often have a sense of purpose centred on care for one another and for their environment. The themes of civic action and migration introduced in this chapter pave the way for the following exploration of *ikigai* (sense of purpose in life) in Chapter 8, as I examine how people strive to 'live well' in later life.

Notes

1. Ohara 2021.
2. Kavedžija 2018.
3. Benedict 2005; Nakane 1970; Doi 1971; Bachnik 1992; Hendry 1993.
4. Kavedžija 2018, 155.
5. The extensive study of migration in the social sciences is part of the so-called 'mobility turn' (Sheller and Urry 2006; Janoschka and Haas 2013; Klien 2015), including research on the experiences of retiree migrants (Ono 2008) and the migration of young lifestyle migrants (Klien 2016) in Japan.
6. Traphagan and Knight 2003, 13.
7. Feldhoff 2020.
8. Benson and O'Reilly 2009.
9. Klien 2016.
10. Traphagan and Knight 2003, 13.
11. Kudo and Yarime 2013.
12. Sakuno 2006.
13. Champion and Shepherd 2006.
14. Ohno 2005.
15. Suzuki 2018.
16. Qu et al. 2020.
17. Klien 2019.
18. Matanle and Rausch 2011.
19. Klien 2016.
20. Nakamura 2020.
21. Matanle and Rausch 2011.
22. Matanle and Rausch 2011.
23. Daniels 2010.
24. Love 2013, 116.
25. Klien 2020.
26. Robertson 1991; Ivy 1995.
27. Smith 1978.
28. Anthropologist Jamie Coates (2019) provides a helpful discussion of 'belonging', including summaries of contributions by anthropologist Michael Jackson (2005).
29. Jackson 2000, 214, cited in Coates 2019.
30. Jamie Coates, drawing on the work of anthropologists Tim Ingold (2000) and Michael Jackson (2000), argues that in the context of mobile imaginaries, a sense of 'being at home in the world' manifests in the alignment between one's actions and personal schemas of belonging (Coates 2019, 471).
31. Mizukoshi et al. 2016.
32. Knight 2003.
33. Benson and O'Reilly 2009.
34. Klien 2015, 2016, 2019, 2020.
35. Knight 2003.
36. McIntyre 2009.
37. Hoey 2010; Korpela 2014; Leivestad 2020.
38. Benson and O'Reilly 2009.
39. Daniels 2010.
40. Miller 2016.
41. Turkle 2012.
42. Traphagan 2004.
43. Klien 2016; Benson 2013.
44. Klien 2016.
45. Rosenberger 2017.

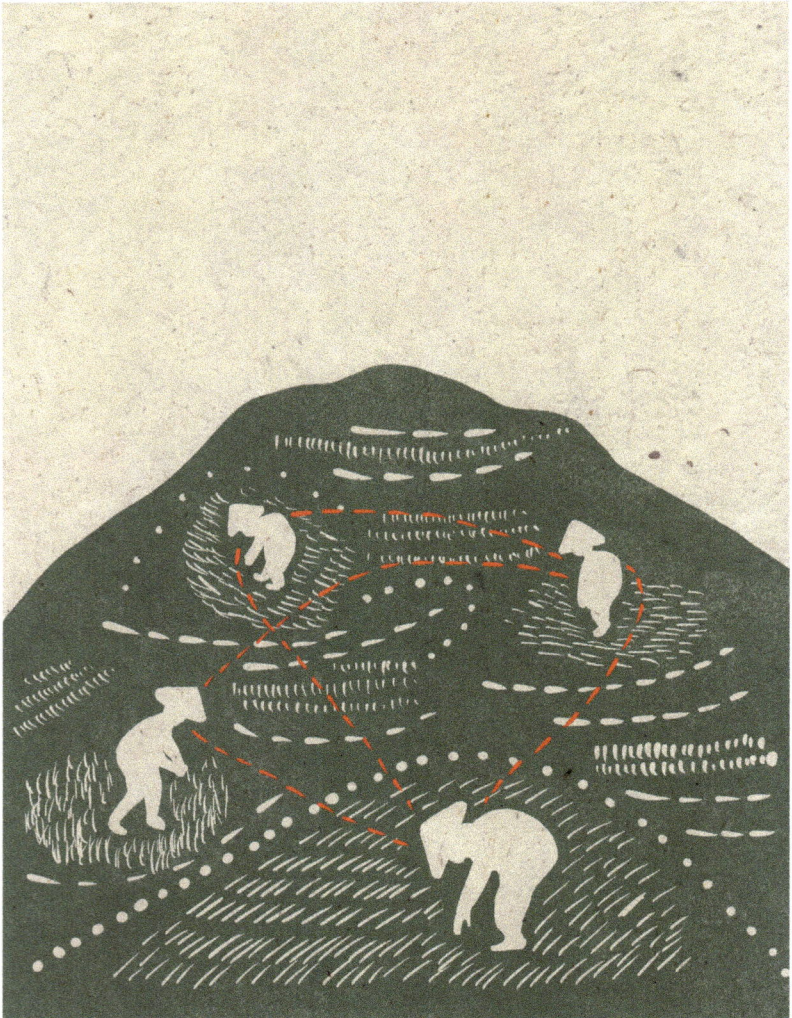

Image by Laura Haapio-Kirk.

8
Purpose in life: *'ikigai'* through the life course

Probably *ikigai* has something to do with longevity. We live longer nowadays after retirement. It is necessary to think about *ikigai*. My parents' generation didn't worry about that. They were only working throughout their lives. If we stop working at 65, there are 15 or 20 years left. So it is necessary to think about how to spend that time.

Kenji-san, aged 68, Kyoto.

Ikigai is a common term in Japan, created from the combination of the words 'to live' (*ikiru*) and 'worth' (*kai*). While there is no direct translation into English, it can be understood as 'that which makes life worth living'. It is thus a concept which has cross-cultural application and meaning outside of Japan, broadly capturing the concept of purpose or meaning in life.[1] Kenji-san's definition of *ikigai*, expressed in the quote above, reveals shifting life courses and expectations between generations. His personal understanding of *ikigai* reflects his wider concern for health, demonstrated through a commitment to regular exercise and self-tracking described in Chapter 6. Through a series of portraits of individuals, in this chapter I show how purpose in life is conceptualised differently by men and woman and by older and younger generations. *Ikigai* was considered a necessity for a healthy later life in general terms by everyone in this research. However, the pressure to 'find' one's purpose appears to increase with age, along with wider narratives of an older individual's responsibility towards society.

'Meaning in life' as a concept is difficult to define. Researchers, primarily from psychology, have identified three facets: coherence (when one's life makes sense), purpose (having core goals) and significance (a sense of life's inherent value).[2] In Japan having a sense of *ikigai* is encouraged at any age, but previous anthropological research has shown that it is inherently associated with later life, when other forms of social belonging such as work or caring for family may fade.[3] My research demonstrates that those in mid-life who are typically still embedded in defined social roles may seek other ways to define meaning and purpose in life centred on narrative-making.

I argue in this chapter that people in mid to later life have a contentious relationship with *ikigai*. Some people reject the term in favour of other ways of expressing meaning and purpose in this life stage, while others repurpose it for their own needs. Just as terms such as 'successful ageing' can provoke an uneasiness centred on questions of what 'unsuccessful ageing' might look like,[4] *ikigai* is a concept that people question and find their own alternatives to – yet one that remains an important component in the way they think about their lives and ageing.

The concept of *ikigai* has been productive for anthropologists seeking to understand how Japanese people construct 'selves' within society.[5] It provides a lens useful for understanding the link between what it means to 'live well' and to have a purpose in life.[6] The 'boom' in popular discourse around *ikigai* in Japan started in the post-war economic growth period, prompted by rising longevity and uncertainty over social roles determined by lifelong commitment to one's company or family.[7] As discussed in Chapter 3, retirement can be particularly difficult for men who lose their social identity and are in danger of becoming 'oversized garbage' (*sodai gomi*) or 'wet autumn leaves' (*nure ochiba*).[8]

The topics of wellbeing, happiness and ideals of the 'good life', while generally receiving limited explicit examination by anthropologists,[9] have received more scholarly interest in anthropological studies of Japan – perhaps because of the cultural prominence of the concept of *ikigai*.[10] In the latter part of the chapter, I draw inspiration from the anthropologist Iza Kavedžija's life course approach.[11] In her work with older Japanese people in Osaka she demonstrates through the life stories how people craft 'meaningful lives' through their efforts to 'live well' in everyday life. I explore how narrative and meaning in life is constructed, in this case through a graphic anthropological approach that seeks to provide a non-linear method of exploring how people create meaning. I then expand my analysis to include the sharing of visual and textual media online as a meaning-making practice, via a discussion of a painting by one of my interlocutors and her social media posts. To begin, however, it is important to understand how *ikigai* has become central to narratives of ageing 'well' in Japan.

'Ageing well' with *ikigai*

Ikigai has been argued to be indispensable for understanding Japanese conceptions of wellbeing.[12] It is a familiar concept in public discourse, often used in relation to the welfare policies and community development associated with Japan's ageing society.[13] *Ikigai* is closely linked to psychological wellbeing;[14] an absence of *ikigai* is even considered to put a person at greater risk of mortality.[15] *Ikigai* was a central component of the Japanese government's 'Active 80 Health Plan', launched in 1999, to encourage a healthy lifestyle among older adults.[16] It is therefore heavily associated with official discourses around healthy ageing and has been proposed as the Japanese equivalent to strategies for 'successful' ageing.[17]

Kyoto City government, for example, has a 'Citizen's Health and Longevity Plan', at the core of which are '*Ikigai* Measures'. The city supports facilities such as 'elderly club houses' and activities 'for creating purpose in life'. Welfare policy encourages older adults to help themselves and each other through volunteering, with the government subsidising expenses for support activities undertaken by local residents for the 'small troubles' of the elderly.[18] Many people I met had absorbed the national and regional policy discourse on *ikigai* and were, in their own various ways, taking it seriously, both as an individual endeavour and as a commitment towards their own future.

Kenji-san, introduced at the start of this chapter, is an avid adult learner. He spends his days at his local university library, for which he has a visiting membership. Accompanying Kenji-san to the library one day, I noticed that several other men his age and older were sitting at desks reading. A few were also sleeping. Kenji-san explained his daily visits to the library as part of his strategy for *ikigai*, something that he felt would keep him in good health:

> *Ikigai* is one of the most important parts of how to live longer in a good condition. If we have no aim in life, we will get older more easily. There are a wide variety of things to do that can be *ikigai*. Some people can enlarge their hobbies or have more contact with society. Without life purpose, people just live. They just spend meaningless time at home. That leads to isolation. If you have life purpose you will become happy and energetic which might be good for longevity.

Kenji-san's association of *ikigai* with a strategy for longevity reflects wider discourses that abound in Japan's ageing society. His ideas also illustrate the social and moral obligations around health in later life that are intertwined with personal responsibility and notions of being a 'good person'.[19] In Kenji's appraisal, *ikigai* is necessary when one reaches later life. This puts it in alignment with wider discourses around 'successful ageing' found in many countries today.[20] Cross-culturally, eudemonic wellbeing (a sense of purpose or meaning in life) is associated with lower death rates.[21] A common thread running through many of my conversations was that it is possible to progress through life without feeling that one is ageing at all if one pursues activities with 'passion' and 'aspiration'. Midori-san, the wood worker and farmer from Tosa-chō introduced in Chapter 7, explained that despite being in his seventies he does not feel that he has changed since his younger years because of an enduring passion for his work:

> I never feel that I've aged. Your physical appearance might change, but the drive you have to achieve your dreams or things that you feel you need to do will never change. Farmers' bodies may get old, but their passion for their jobs has not changed since they were 20! Whether you find your *ikigai* at 20, 30, 40, you will never age if you have that passion. Never. If you are young and have no passion (*jōnetsu*) or aspiration (*ganbō*), you are mentally and spiritually ageing.

Here we see how *ikigai* is conflated with 'passion' (*jōnetsu*) or 'aspiration' (*ganbō)* through work. However, lifetime employment that might have once fulfilled a sense of *ikigai* for older generations no longer holds similar promise for the young. Young people in both my urban and rural fieldsites told me of the pressure they felt to find their *ikigai*, with teachers stressing the importance of finding a career to dedicate their life to, despite the realities of labour precarity and temporary contracts. In being urged to seek work they are passionate about, and thus to follow their *ikigai*, despite worsening economic and labour contitions, there is a parallel with ideologies about 'passionate work' in which self-fulfilment is prioritised over economic security.[22] Anthropologist Emma Cook's work with 'freeters', young part-time workers aged 15 to 34, shows how they epitomise contemporary structural precarity, shaped by shifts in the Japanese employment landscape since the 1990s recession.[23] She argues that youth employment in Japan, and globally, faces heightened precariousness compared to previous generations, necessitating adaptation to a markedly transformed job market.

Cultural theorist Lauren Berlant proposes the concept of 'cruel optimism', highlighting the inherent disappointment and dissatisfaction that arise when unattainable or harmful cultural and social ideals shape desires and aspirations.[24] By examining critically the idealised notion of 'the good life', particularly in relation to labour, the family wage and aspirations for upward mobility, Berlant suggests the possibility of envisioning alternative ways of living and engaging with the world that are less dependent on unsustainable ideals. Although Berlant's analysis primarily focuses on an American post-Fordist context, there are notable parallels with the contemporary situation I observed in Japan, for example in the fragmentation of labour as discussed in Chapter 3.

However, many of the older people I met during my fieldwork, particularly in rural Tosa-chō, shared a sense of self-fulfilment that was grounded in the present, in the act of cultivation, rather than future aspirations. Like Midori-san, many people continued to work their land well past the official age of retirement, attributing this activity as a source of their vigour and indeed their *ikigai*. This purpose in life was very much rooted in the daily activities that make up the present. As Amrith, Sakti and Sampaio articulate in their introduction to their edited volume *Aspiring in Later Life*, aspiration is not always future-oriented, despite much scholarship presenting it as such.[25] In fact, for Midori-san, his idea of aspiration can be understood as oriented towards his younger self, connecting him with a lifelong passion that continues to sustain him.

This was a common understanding of *ikigai* among many people I met. The ability to wake up in the morning and know what one needs to do, whether for a hobby or for work, was the sense of purpose thought to keep people connected with their youth. As Midori-san succinctly put it. 'You get old and have no passion? You'll age.' Midori-san elaborates further on his thoughts about *ikigai* in the video below (Fig. 8.1).

Previous ethnographic research among older adults in Japan has shown how 'meaning' in life can be created through various practices, such as making an effort to keep healthy and active, to do things the right way (*chanto suru*)[26] and to cultivate a general 'disposition of care'[27] which situates people in a network of reciprocal concern for others. As discussed in Chapter 7 on health and care, many participants were motivated to stay healthy because they did not want to burden anyone else in their old age. Warding off dementia by staying active was thought to show consideration for others by not becoming a burden or 'troublesome' (*meiwaku*).

However, anthropological work in Japan has shown how dependence and independence are not binary opposites – they cannot in fact exist without each other.[28] As Takie Lebra outlines, binaries are thought of as a sort of 'oppositional logic' in Western/Enlightenment thought. In the Japanese case, however, supposed binaries such as dependence and

Figure 8.1   A short video about *ikigai* as finding work to dedicate one's life to. https://youtu.be/FJlbRvn6tLU.

independence can be better understood as intertwined and relational.[29] For example, the concept of *amae*, or dependence, involves the presumption that one's needs will be indulged by another, as famously written about by psychoanalyst Takeo Doi.[30] *Amaeru* (to depend) is thus contingent on *amayakasu* (to indulge), carrying a flow of emotion, love or affect. In this cultural context Lebra states that the question of autonomy is unlikely to arise.[31] I found that being depended upon by others was the source of many people's *ikigai*, which in turn they themselves depended on for happiness.

Historically the meaning of *ikigai* was based on the social value of a person's life. However, this meaning has now shifted to incorporate the more individual values of happiness and fulfilment.[32] People often expressed both inflections of the term and believed that they were implicated in one another. Through the following example, we see how the health benefits of having responsibility and being relied upon through volunteering far outweigh a desire to enjoy an 'easy' retirement. Having something to get up for in the morning was regarded by Katsumi-san and Takao-san, both men in their seventies in Tosa-chō, as crucial for health in later life, including for warding off dementia. These two men had been elected as local leaders for their village. For the past six years they had been engaged in various community-building initiatives, such as redeveloping an abandoned school into a community centre and establishing a Sunday market where locals could sell their produce and also meet one another for lunch. Sitting in their office in the school, they explained how important their work was for their own sense of wellbeing and happiness, despite initially having other plans for their retirement:

Katsumi-san:	I wanted to spend my retirement nice and easy (*yukkuri to*), but we were both asked to do so many things! So no time to rest!
Laura:	Did you feel the responsibility to protect your village?
Katsumi-san:	Of course! It's not like you get paid for this job so you need to feel that responsibility (*sekinin*).
Laura:	Does that make you feel happier?
Katsumi-san:	How are you supposed to do it without the feeling of happiness (*shiawase*)?
Takao-san:	I think in one's life, being relied on (*tayora reru*) is the most important factor for happiness. You are busy, but that gives you happiness.
Katsumi-san:	You feel a sense of accomplishment (*tassei-kan*).
...	
Takao-san:	This job, working, is my fuel for happiness.
Katsumi-san:	I will get dementia soon.
...	
Takao-san:	But you won't get dementia if you work!
Katsumi-san:	There is nothing sadder than waking up in the morning and not having anything to do.
Takao-san:	I think I really don't want to go to the hospital. I really don't want to die at the hospital. I want that for everyone.
Laura:	How would everyone achieve that?
Takao-san:	Through work, or pursuing a hobby.

This discussion makes clear how much caring for the community is thought to keep one healthy; by staying healthy, one takes care of the community. However, Takao-san also had a strong sense of individual responsibility:

> When you get older it's most important to take care of yourself. Not to fall or trip, take your medicine. You know your body best, so caring for yourself is so important. Everyone has to work and do that for themselves.

Work was a primary way of maintaining their responsibility towards their own health. Both Takao-san and Katsumi-san had previously been employed in stable, long-term jobs at the Post Office and with the Japan Agricultural Association respectively. Their current roles were unrelated to their previous work; they were simply called upon to become community leaders and stepped up to the task, thereby maintaining their sense of *ikigai*.[33]

Chapter 3 showed how Silver Human Resource Centres (SHRC) attract older workers by framing their labour as virtuous participation in society, no matter the type of work involved. Studies have demonstrated the link between working through a SHRC and feelings of *ikigai*.[34] Such centres were set up to enable older people to maintain an active role in society by seeking part-time employment and were explicitly designed provide an opportunity to develop *ikigai*.[35] For many people, especially men, the social role of 'worker' was more valuable to people's sense of themselves than the kind of work performed. Due to lengthening lifespans, upon retirement people are spending more time outside of socially scripted roles that positioned men and women as workers or domestic labourers respectively. Kavedžija asserts that in such a context, '*ikigai* thus becomes a social issue, a matter of concern, when it no longer "naturally" unfolds from the social role itself'.[36] The disintegration of cultural scripts for the position of older people in society has put older men and women themselves at the forefront of defining their own ways to age 'well'.[37] In the next section I will examine how gendered differences affect notions of *ikigai*.

Gendered *ikigai*

For men, continuing work in any capacity tended to be associated with *ikigai*. For women, however, the notion of *ikigai* associated with work tended to be more aspirational. It was about switching social roles after being care givers towards shaping their identity around their unfulfilled dreams, for example travelling or starting their own business. *Ikigai* can also refer to self-realisation in post-retirement life for men after being released from work-related hierarchies of value through separate activities such as retirement migration.[38] For Kenji-san, the adult learner introduced above, an early retirement at 58 meant that he could finally pursue a university education and take up a Master's course abroad – something that he had delayed because he had 'good working conditions', meaning a comfortable salary and a steady job at the same telecommunications company throughout his working life:

> When I was working I didn't think about my future. But when I stopped working I started thinking about studying, that became

my *ikigai*. Probably I always wanted to study abroad, but I couldn't stop working because my working conditions were really good. But I should have stopped working earlier. I would have more opportunities to fulfil my goals. When people have good conditions, they hesitate to get out of those good conditions.

For Kenji-san, who recognised the disparity between his own 'good conditions' and those of others in more precarious employment, his fortunate situation was also a kind of trap. Yet the freedoms afforded to him as a single man, with no dependents, meant that he could take the risk of early retirement. With the government effectively raising the retirement age to 70 years old, as discussed in Chapter 3, and with increasing financial precarity facing workers of all ages, Kenji-san's choice to retire early to pursue his interests may not be one available to many others. He also felt that married men were more burdened by this change of lifestyle because of its greater impact on their domestic life:

> At the time of retirement, men want to continue working because working is a type of *ikigai*. In most cases when wives have a retired husband, they don't want them to be at home all the time. So it is better for their wives if they are out and in good spirits. For housewives, their *ikigai* is to get divorced [laughs].

For older men in particular, the transition from working life to retirement can be a time that alters their identity, because their notions of self are so closely tied up with work, as outlined in Chapters 3 and 4.

For some men, such as Tadashi-san, a 'U-turner' who had moved back to Tosa-chō in his sixties with his wife, getting older (he is now in his late seventies) meant that he was able fully to embrace a new identity through a different kind of work which he had been developing for some time. In his forties he put his savings into setting up a side-business growing flowers cultivated from cuttings of mountain plants. Now he was able to work full-time for this business, after retiring at 65 from being a taxi driver in Kōchi city.

> When you reach a certain age, you want to do whatever you want to do. You just want to have fun! You're working so hard up until then, right? … I had a strong passion for flowers and gardening and I wanted to live it (*ikitakatta*).

This passion was largely what prompted Tadashi-san and his wife to move back to his rural hometown.

This decision posed a problem for his wife Misako-san, however, as the connection to her friends and their children and grandchildren who remained in the city was what gave meaning to her life. Misako-san's experience of moving from the city to a place where she did not know many people was not easy and she subsequently suffered from depression. She struggled with the remoteness of their mountain home, far away from her friends in the city. Her face lit up when she brought out her smartphone and showed me photos of her grandchildren in their family LINE group. She was also glad to be part of a LINE group for a flower club for local people who were also interested in growing flowers, and the group would arrange regular exhibitions of their flowers together.

While their home was at the top of a winding mountain road, they were able to stay connected through their smartphone that they shared. Misako-san, in particular, valued the digital connection to maintain her happiness. She also took great comfort in her dog and the wildlife of the area. Flicking through a physical photo album dedicated to a flying squirrel that she had rescued and rehabilitated, it was obvious how caring for others, both human and non-human, was central to the way in which she attached meaning to experiences in her life. While wider discourses of *ikigai* that emphasise activity and maintaining passions in life were repeated by my interlocutors, my ethnography revealed that being embedded in relationships of care continued to be how meaning and purpose in life manifested, especially for women.

One day, while I was attending the bi-weekly lunch in Tosa-chō, intended as an event where elderly people and young families can mix, the woman from the town hall who co-ordinated the activity with the older volunteers explained:

> We work in welfare because we believe happiness exists because everyone has a reason why they're here. Babies, children, everyone. Children create *ikigai* for us, you know? Everyone has a reason and that's crucial in life. So people in their eighties create *ikigai* not only for themselves, by helping at this mothers' lunch, but also for us younger people because we care for them. So the most crucial aspect of happiness is to know the reason why you exist and your purpose.

All the older volunteers at the event were female, as were most of the attending parents of the young children. Being a volunteer helped women in a similar situation to Misako-san, in that their *ikigai* had been bound up with work and caring for a family. Such traditional gender roles might be limited now that children were adult and grandchildren were often living remotely.

I found that some people maintained a multifaceted understanding of *ikigai* that reflected changing social scripts, which for women often included their care roles. As Kazuko-san, a woman in her early sixties from Kyoto, explained:

> You know, there are several factors for *ikigai*: one is pleasure and another is purpose and obligation. I think *ikigai* consists of these factors. Now I have so many responsibilities, they increase with age, including taking care of things in my neighbourhood. Maybe… taking care of my mother-in-law is now my *ikigai* because right now it is my obligation (*gimu*). This is a factor of my life. My reason to live.

Kazuko-san's relationship with her mother-in-law was strained. She had many responsibilities because of her husband's position within the family: 'It's hard to be the wife of a first-born son!' Yet despite the hard work, being a person on whom others depend and fulfilling one's social obligations were two of Kazuko-san's 'reasons to live', along with the pleasures of seeing her favourite band and updating her blog with reviews of the latest concerts she had attended. The relationships she developed through her activities in the neighbourhood also enmeshed her in reciprocal care, along with her friendships with women of a similar age. Care thus situates people within webs of relations with other people, which is at the heart of crafting a meaningful life.[39]

Generational *ikigai*: from relation to aspiration

These ethnographic observations confirmed the way in which *ikigai* today has quite a variety of meanings and interpretations, but often still diverges along gender lines. I subsequently decided to employ graphic methods to examine the various representations of the concept visually; these then revealed a further line of division between generations. The images below (Fig. 8.2) show drawings made by my interlocutors, visually depicting their understanding of *ikigai*.

The drawing on the left was created by an 18-year-old and the one on the right by a 70-year-old. Both of them were women living in Tosa-chō. The one on the left was typical for all the younger and middle-aged people I conducted this exercise with (nine). It showed an individual at different stages in their life, typically ascending a series of steps, leading

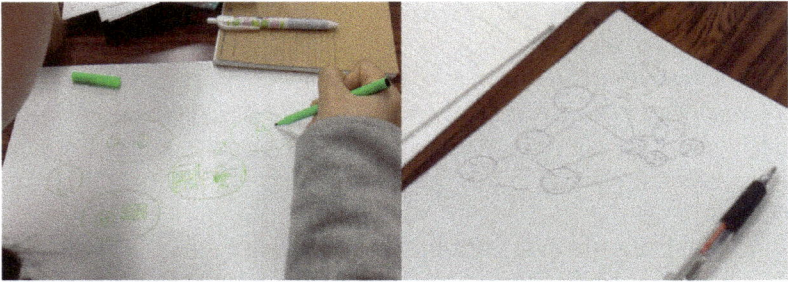

Figure 8.2 Drawings made by participants about their *ikigai*. Images by anonymous research participants.

to a certain goal or dream that they had since they were young. The image on the right was striking because of its unusualness. Instead of a progressive series of steps, Yamada-san drew herself as connected with others in a web of smiling faces.

At the age of 70, Yamada-san was precisely positioned between the older generation, whom she felt had very traditional ideas about *ikigai* centred on social value, and the younger generation, who tended to be more aspirational. Yamada-san would always bring delicious home-made sweets to social gatherings because she knew that these would make people happy. She shared the same conviction of the previous generation that *ikigai* can be found in service to others:

> My mother-in-law used to plant rice and tell me how exciting it is to think about sharing our rice with everyone. I realised that my mother-in-law's *ikigai* was to share this rice with everyone and see their happy faces. That song, 'you raise me up to climb the highest mountain …', you know that song? It's kind of like that! Making other people happy lifts her spirits up and makes her happy too. And then when I think about my *ikigai*, it's to live a happy life and to make everyone happy too. Thinking about how everyone can be a little happier than the day before … I like that and I think that might be my *ikigai*.

Yamada-san, like the generation before her, situated herself within a web of social relations that give her a sense of *ikigai*. However, Yamada-san was also able to see how her children viewed their own life purpose differently. It was through raising her family that she realised that she did not have to follow the same path as previous generations of women for whom *ikigai* could be restricted:

The word itself is very heavy ... it makes it really serious. I don't usually use this word. *Ikigai* feels really restricted. It feels claustrophobic. 'I want to live this way' (*ikitai kō yatte*) that feels more free, like freedom! *Ikigai* is not freedom. Old Japanese people, especially old women, just got married and had no freedom. They just lived. A woman works for her husband and husband's parents until she dies. Until she dies, a woman doesn't have freedom. She keeps her thoughts in. I have that image. *Ikigai* is a bad word ... that's the image I have.

Yamada-san felt that *ikigai* for women in the past was solely constrained to domestic labour. At the same time, however, she felt that her conception of *ikigai* expanded while raising her children:

I've lived my life thinking there is a right way to live. When I had children, I realised my children had a whole new view of the world than I did. The way they saw the world was 180 degrees different from me! I realised it didn't have to be a certain way. I feel this way, but you feel another. And that's OK. While raising my children I was meeting different people. Other mothers, children's friends, teachers, a new environment ... But also going out myself and meeting new people. My idea expanded. Ohhh that might be an *ikigai*! And having that new perspective meant I wasn't strapped up in my home any more.

Through raising her children, Yamada-san gained a new sense of the possibilities that *ikigai* could entail. However, it was clear that she was still ambivalent about the term and was in the process of defining it for herself.

Something you cannot live without

There were others who felt that they had not found their passion yet. They felt anxious that they should find aspirational goals and were reluctant to use the term *ikigai* to describe other kinds of activities that kept them busy. Sayoko-san, a woman in her sixties living in Kyoto, retired from the job she had been doing for 30 years and then started up several of her own businesses. However, she would not consider these to be her *ikigai*; they were just a way for her to earn money after she and her husband divorced. She felt anxious that time was running out and she must find a new central focus in her life:

> I'm still searching for my *ikigai* right now. I worked so hard until I was 50 and I was also taking care of my kids. When I was your age [early thirties], I told myself I was going to retire from teaching at the age of 50. And when I did, I started my new business. Last year [at the age of 58] in some sense I retired again … I started a guest house and then this year, I'm still thinking what my goal is. I try to have a simple life now, and do some volunteer work, teaching English for free … and now my mother is in a care home … My goal is to go to school again, to a language school maybe in Europe, but I need money. Financially, I don't have enough money to live abroad. Without money I can't do anything, I got divorced so I need to take care of myself. The decade from 60 to 70 I think is going to go so quick! So I have to rush!

There was a sense among many of my recently retired research participants that finding one's *ikigai* had become imperative to staying healthy, as the individuals earlier in this chapter demonstrated. They also shared a keen sense of time passing, as the next case will highlight.

I first met Megumi-san at the bi-weekly parents and volunteer lunch in Tosa-chō, as mentioned above. She was attending with her three-year-old daughter who had been born when Megumi-san was 43, after she had moved with her two older daughters and husband from Tokyo to Tosa-chō following the triple disaster of 3.11. In her twenties Megumi-san had trained as a painter at art school, and she had subsequently had three solo exhibitions. During my fieldwork Megumi-san often spoke about her desire to resume her painting career, something that she had put on hold while raising a family. As she explained,

> I chose this way. I didn't want to stop drawing, but I wanted to have another dream with my husband. So, I followed his dream.

She had supported her husband with his ambition to open bakeries in Tokyo. This endeavour had been highly successful, with media attention garnered through Megumi-san's social media activities. Upon their move, the couple established a popular bakery and cafe in their new rural home. The family enjoyed life in Tosa-chō and Megumi-san felt that in this rural environment she was able to gain inspiration for her artwork and time for herself, even with a young child to look after.

Tragically Megumi-san's husband passed away suddenly towards the end of my fieldwork. In the two years after her husband's passing Megumi-san had come to a form of acceptance of the situation, along with a realisation that she needed to focus on her *ikigai* as a way forward:

> We seemed to be living a rich rural life. However, my life has changed after my spouse's sudden passing. His life ended the morning after my birthday and ever since pain and regrets never leave me. I have to live every day around it, telling myself that it's OK to find my own happiness. It seems my life has been a chain of unexpected events. I chose the way not knowing where the road leads me. It's been an eventful journey. But one thing I am sure about is that I want to express myself through art, and this I call *ikigai*. The importance of one's life, something that you can't live without.

Two years after returning from Japan, I invited Megumi-san to produce a painting on the theme of *ikigai*, which we then discussed via video calls. I was reassured when she told me that the timing for such an invitation was just right; she was ready to start putting more energy into her artwork after such a life-changing tragedy. The resulting painting (Fig. 8.3) depicts her life course as she approached the age of 50. It captures how she sees her life as composed of contrasts: day and night, mountain and river, countryside and city life, sorrow and joy. The painting also depicts the motivation behind the move that her family made from Tokyo to Tosa-chō. Megumi-san uses traditional Japanese materials such as *washi*

Figure 8.3 Megumi-san's painting expressing her concept of *ikigai*. Image by Ito Megumi.

paper, handmade brushes, natural pigments and glue to make her paintings. The pigments are the same ones that she used 25 years ago when starting out on her artistic journey; the are derived from natural materials, including oyster shells and minerals. Using these materials is important to her practice, connecting Megumi-san to the craftsmanship that goes into their production which she feels celebrates precious knowledge and skills, and also connecting her with her past.

The painting was then turned into a multimodal artwork embedded with audioclips of Megumi-san discussing each element, published elsewhere.[40] The intention was to invite the audience to explore Megumi-san's life story in a non-linear way, exploring connections between different visual and narrative elements as Megumi does when thinking about the interconnectedness of life events across time and space.[41] Her narrative situates her early interest in painting rice fields as a form of destiny-making that connects with her present life in the countryside. At the start of her artistic journey, she remembers being fascinated by photos of small and irregular rice fields in a photo book that she purchased; this prompted her artistic focus of landscape painting. Looking at the book again after 25 years, she realised that one of the photos had been taken close to her current house in Tosa-chō.

> It gives me goosebumps. I was so surprised to see, and at the same time I was destined to be here (*unmei-teki*), I thought.

By focusing on Megumi-san's visual expression and enabling the audience to hear her speak about her life, a multimodal presentation can help to convey the same sense of interconnectedness and meaning that Megumi experiences when thinking about her life journey. The image presents a narrative with no beginning, middle or end. In so doing it aims to capture the sense of wonder that Megumi-san feels at how the 'turning points' in her life, as she calls them, interconnect.[42]

Experiences that Megumi-san previously took for granted surface as the most meaningful when seen from a more distant vantage point, one in which everything around her has changed. Looking back at the period when the family had just moved to Tosa-chō and were enjoying a happy family life, Megumi-san explained that it is only now that she is able to see the significance of that time during which they did not know 'the deep meaning of life. It's happiness, but I didn't realise that this is the moment of our life happening just one time'. Working her way through grief, she has come to find meaning even in the most difficult parts of her life.

Psychiatrists have identified meaning-making as an integral part of the grieving process.[43] Anthropological work has also demonstrated the ways in which loss is not exclusively associated with decline, but also has meaningful value.[44] The construction of her painting around a series of contrasts reflects her feeling that when bad things happen in life, these can also be opportunities for personal change and even growth. She naturally has bad days, but she tries to catch herself and turn her thinking around. Now that she is almost 50 years old she wonders how many years she will have left. Megumi-san was almost the age of her husband when he died – their birthdays being just four days apart.

> His life ended but my life kept on going. I think I have started counting from the day after his death, so now I know, I really know, that there will be an end for sure. Yes, that's really changed; from that moment I changed my way of thinking for myself and for everything surrounding me.

Megumi-san's sense that the clock is ticking prompts her to ask herself how she wants to live. Having a sense of *ikigai*, or purpose in life, is for Megumi-san part of her practice of narrative-making, prompted by life-changing events. She has become more motivated explicitly to pursue *ikigai* because of the realisation of the ephemerality of life, something that many of my older research participants shared. Megumi-san has reached this observation perhaps earlier than others because of her experience of loss.

Megumi-san's visibility to others around her, especially to her own children, motivated her to 'live selfishly', in her words, by living her life in pursuit of *ikigai* and thinking about what she wants to do in her remaining years.

> *Ikigai* to me is something that I love to do, but also it needs lots and lots of effort to keep it on my mind every day … When I was young, I wouldn't think this way.

This effort to pursue *ikigai* is socially motivated and is reflected also in the way that she uses social media. After her husband's death, Megumi-san received support from friends and acquaintances via Facebook posts that she would occasionally share (Figs 8.4 and 8.5). She would polish the words of each post, which would tend to be quite open about her journey through grief, but would also focus upon the positive things in her life. Megumi-san received many public and private comments that

Figure 8.4 A fiftieth birthday Facebook post by Megumi-san. Screenshot by Laura Haapio-Kirk.

Birthday weekend
Memorable half-century birthday & a day to remember a sad farewell

Thank you for giving me your heart

Surrounded by flowers I love
Laugh a lot and cry
Eating & drinking rice & cake made by my loved ones
Reset my mind!
At the end of the day, my crush is also charged at the recommended online concert lol

I got power within my body

Thank you so much
I'm going for a slow walk again~

#50thbirthdaycelebration #deathanniversary #sadness #happiness

Figure 8.5 A gardening Facebook post by Megumi-san. Screenshot by Laura Haapio-Kirk.

It really helps if it's sunny on the weekend 😊
Finally, the flowerbed is completed.

Waiting for half a year after asking a carpenter for a fence...
Making full use of wonderful craftsmanship, a wall made of local cedar wood is completed!
Just in time to stake the roses.

Stack the surplus block walls with mortar
↓
Paste sliced bricks with cement bond
↓
Fill with white mortar
↓
I pasted the floorboard waste material on the top.
(Mostly by friends, I'm an assistant)

The memorial fig tree for my husband, which had died due to transplantation and heavy pruning, has also revived.
The vitality of plants... it's amazing.

thanked her for being open in sharing her experiences. She was initially surprised to find that so many people were viewing and responding to her posts. This visibility was not new, as she had previously often posted about their family bakeries. However, her social media use was now far more personal and centred on her own experiences. When she realised that people were watching, she felt a desire to demonstrate to others that she and her family were doing OK.

Megumi-san is motivated to live well and try to thrive despite adversity, both for her own sense of living with *ikigai* and to demonstrate to those around her that it is possible for life to continue even in the most difficult times. Visibility can thus be understood as connected to *ikigai*, which links with the earlier discussion on the original social nature of the term. Having *ikigai* was traditionally linked with having social value.[45] Therefore in displaying to others that life can be meaningful despite personal tragedy, and in encouraging others to be open in talking about difficult things in their life, Megumi-san is seeking to enrich her community by setting a positive example. Much like Megumi-san's paintings, which are a material manifestation of her *ikigai*, her social media posts are also ways in which she makes visible her efforts to live with purpose. She shares updates on home and garden renovation projects (Fig. 8.5), for example, and comments on how opposite feelings of happiness and sadness are combined, especially at important moments in her life (Fig. 8.4). Megumi was thinking about this particular balance of antithetical feelings when painting. She drew associations with the concepts of yin and yang from ancient Chinese philosophy to depict happiness and loneliness, light and dark.

> Even though you think it's a bad thing happening to you right now, it's not a bad thing. That's just only one side. After several years, after some time, you realise that that was the turning point of your life to change to a better side, or to find a different self.

Anthropologists have shown how moments of personal transformation occur in times of liminality[46] and that 'selves' are subject to change.[47] The 'turning points' referenced in Megumi-san's painting, and in her ongoing activity on social media, can be understood as visual representations of rites of passage that, when shared with others, enable a form of 'communitas'[48] that supports her on her journey.[49] Similarly, Megumi-san's social media posts are a form of visual ritualisation that enables others collectively to participate in her transition from her previous life into her new reality, in which *ikigai* is foregrounded as a way of living.

Anthropologists are increasingly turning their focus to existential matters.[50] In particular, ethnographies that investigate modes of narrative-making, such as storytelling,[51] and now also social media posting,[52] have emerged as key to understanding how people make meaning of their lives. Anthropologist Shireen Walton argues that the smartphone has become an 'existential object' in the way that it 'forms an intimate link between the self and with narrative'.[53] Megumi-san's Facebook posts highlight the more existential side to social media, challenging the dominant critique of superficialness often levelled at it. Instead, we see how constructing narratives around life events on digital platforms can be central to the ways in which people articulate and pursue meaning and purpose in their lives.

Conclusion

This chapter has shown that *ikigai* is far more complex and diverse than captured by popular discourses that frame it only in terms of 'successful ageing'. The meanings of *ikigai* change depending upon age, gender and the economic and social conditions of contemporary Japan. Throughout the chapter, the clearest finding regarding *ikigai* is the imprecise nature of the term. Each individual presented has their own, constantly shifting response to the term and ways of living well, in response to their social and personal obligations and aspirations.

Kenji-san, for example, values the opportunity early retirement has afforded him for higher education. He pursues this aspiration out of his interest and awareness of the value of such activities for longevity. Some people, such as the wood worker Midori-san, claim that *ikigai* means finding work that one is passionate about, which results in one effectively never ageing. Katsumi-san and Takao-san ward off dementia and gain happiness by working for their community. Tadashi-san and his wife Misako-san pursue different ways to find fulfilment, revealing the gendered dimension to *ikigai* which I found tended to be associated with care for women and work for men. However, Kazuko-san demonstrated that multiple understandings can co-exist simultaneously in one person, with *ikigai* for her meaning both personal pleasure and obligations of care. For others the term *ikigai* was experienced as a kind of pressure to find the right aspiration before it was 'too late'. Megumi-san embraced the term *ikigai* in her mid-life after losing her husband. She realised that she needed to commit to pursuing *ikigai* in her daily life to find meaning, and to show to herself and others that she was living well despite adversity.

For my middle-aged interlocutors, the topic of *ikigai* elicited bemused faces and a general sense of disassociation from the term; they did not have to think about it yet since they were not yet 'seniors'. Working long hours and leading busy lives was normal for many, but they would not describe this kind of activity as their *ikigai*. For them the concept had a 'heavier' (*omotai*) meaning – possibly because of the association with discourses in Japan's ageing society that position a lack of *ikigai* among older adults as a cause for concern and institutional intervention.[54] The generational difference between relational and aspirational understandings of the term align with the historic association of *ikigai* with social value, something that has now shifted towards more individualised notions of happiness and aspiration.[55] However, I have shown how such notions are intertwined and can be equally motivating as a sense of 'purpose in life'.[56]

As an alternative to 'successful ageing',[57] *ikigai* provokes its own set of conflicting responses. Rather than identifying with the term *ikigai*, 'aspiration' and 'passion' emerge as more closely aligned to what drives my middle-aged participants to maintain active engagement with the world. For many, the term *ikigai* was rejected as being inherently associated with old age, a time when other forms of social belonging such as work or caring for family may fade. For those in mid-life who are often still embedded in defined social roles, *ikigai* appears less relevant. However, we saw how turbulent life events such as divorce or the death of a partner can prompt a reassessment of identity leading to narrative-making practices both online and offline. In Megumi-san's embracing of the concept *ikigai* as part of how she makes sense of her life course, we learn how moments of upheaval can catalyse profound shifts in identity, offering opportunities for personal growth and self-discovery amid the complexities of mid-life transitions.

Notes

1. Mathews 1996a and b.
2. Martela and Steger 2016.
3. Kavedžija 2019.
4. Rubinstein and de Medeiros 2015.
5. Mathews 1996b.
6. Kavedžija 2019.
7. Mathews 1996a.
8. Mathews 1996a, 15.
9. Kavedžija 2019, 201.
10. See for example Mathews 1996a; Mathews 1996b; Kumano 2018; Mathews and Izquierdo 2009; Kavedžija 2019.

11. Kavedžija 2019.
12. Nakanishi 1999; Kumano 2018.
13. Ono 2015.
14. Kumano 2018.
15. Sone et al. 2008; Tanno et al. 2009.
16. Nakanishi 1999.
17. Azuma 1999, 138.
18. Kyoto City official website 2016.
19. Traphagan 2004, 9.
20. Lamb 2014.
21. Steptoe et al. 2015.
22. See Hong 2022. This kind of ideology is prevalent in the creative industries, with some arguing that it is a form of neoliberal subjectivism (Gill 2014; Chia 2019).
23. Cook 2013.
24. Berlant 2011.
25. Amrith et al. 2023. For a thorough overview of the diversity of forms that aspiration may take in later life, see the entire volume.
26. Kavedžija 2019, 80.
27. Kavedžija 2019, 17.
28. Kavedžija 2019.
29. Lebra 2004.
30. Doi 1971.
31. Lebra 2004, 19.
32. Wada 2000, cited in Kavedžija 2019.
33. This finding is in contrast to the findings of Garvey and Miller's ethnography in Ireland, conducted in parallel with this research as part of the ASSA project, where people looked for activities unrelated to their previous employment (Garvey and Miller 2021).
34. Shirai et al. 2006; Weiss et al. 2005.
35. Okochi 1989, 9–10, as cited in Bass 1996, 70.
36. Kavedžija 2019, 3.
37. Kavedžija 2019, 3.
38. Migration may also relieve Japanese women from gender-based caregiving roles (Ono 2015).
39. Kavedžija 2019, 8.
40. Haapio-Kirk and Ito 2022.
41. In a separate publication, I embedded interview clips of Megumi-san discussing her life story, obtained by using her painting as a medium for elicitation during a remote interview conducted online (Haapio-Kirk and Ito 2022).
42. Such multimodal forms created through collaborations between researcher and research participant can also help to decentralise text as a medium of expression in anthropology.
43. Kübler-Ross and Kessler 2005; Neimeyer et al. 2014.
44. Danely 2015a.
45. Wada 2000.
46. Turner 1969.
47. Kondo 1990.
48. Turner 2012.
49. Anthropologist Victor Turner's work in the Congo (Turner 2012) drawing on Arnold van Gennep's work on rites of passage (van Gennep 2019 [1960]) demonstrates how thresholds and liminality are central to ritual practices which have the social function of dispelling conflict.
50. See Jackson 2012, Jackson and Piette 2015.
51. Kavedžija 2019.
52. Walton 2021.
53. Walton 2021, 157.
54. Kavedžija 2019.
55. Wada 2000.
56. Mathews 1996a.
57. Rubinstein and de Medeiros 2015.

Image by Laura Haapio-Kirk.

Image by Laura Haapio-Kirk.

Image by Laura Haapio-Kirk.

Image by Laura Haapio-Kirk.

9
Conclusion

One month before finishing the writing of this book, I received a flurry of messages from Misako-san – one of my research participants in Kyoto, introduced in Chapter 3 along with her mother. Via Facebook Messenger she sent me a series of photos of the beautiful autumn leaves at my local temple close to where I lived in central Kyoto, along with the message 'At Kitano temple now!'[1] The immediacy of photos taken and shared on the spot transported me back to the temple and surrounding streets, as well as all the warm interactions with people there. We later exchanged messages and Misako-san mentioned how her mother was thinking of me and our encounters during their visit, prompting her to send the photos there and then.

A few weeks later I received Misako-san's annual New Year letter in which she shared a round-up of her news. This was a typed letter that she sent to a group of recipients, personalised by a handwritten note on the bottom that read:

> How exciting to stay in touch although we are far away now. Unbelievable! Don't you think? Mother and I thought of you at Kitano temple. It's still a vivid memory for us that you were here or there. We hope to see you again! Let's keep in touch!

In doing so, Misako-san reiterated her digital messages while underlining our connection through the physicality of her handwritten note. As anthropologist Inge Daniels observes, based on her work on New Year cards, the addition of handwritten 'hitokoto' (literally meaning 'one word') messages to mass circulated cards is a common practice in Japan. She argues that this action infuses the card with affective qualities, as these notes serve as a form of 'phatic communication', signifying continued investment in the relationship and imbuing the card with 'heart' (*kokoro komotteiru*).[2] In this book I have argued that visual communication is taking on such affective qualities in the digital sphere.

In the example above it was striking how different the digital and analogue modes of communication were. I was able to feel the connection to place and people through the immediacy of on-the-spot photos shared digitally, whereas the letter was somewhat more formal. In the letter, however, Misako-san was able to go into detail about the renovations she had made to her mother's house. They had turned it into a two-generation home (*nisetai jutaku*) with separate living spaces for herself and her mother, similar to the Katsura residence described in Chapter 2. The renovations had proved difficult, with both women living in Misako-san's small separate flat until the work was complete. However, the divided home was worth the wait. With everything replicated twice, they were now able to live independently, but Misako-san could be on hand if her mother, who was aged 88, needed help. This transformation of the physical home into a space for interdependence mirrors the way in which I found digital technologies are being adopted by older adults to facilitate both independence and connection, closeness and distance.

With older people wishing to reduce their burden on younger generations[3] and working-age people wanting autonomous living arrangements, I found that the ideal of 'ageing in place', i.e. remaining in one's own home in later life, is facilitated through a range of material and digital practices. The smartphone is becoming a space for new forms of multigenerational co-residence and care, paralleling the affordances of two-generation houses. Earlier research on mobile phones in Japan demonstrated their centrality to family relations.[4] but my work offers ethnographic insight at the precise moment when older people were widely taking up the smartphone, opening up a wide range of possibilities that go beyond the capacity of the feature phone (*garakei*).

In this book I have examined the ways in which particular capacities of the smartphone, such as visual communication, intersect with experiences of ageing and practices of care. The aim was to analyse how the recent rise in smartphone adoption among people in later life situates them in Japan's ageing society, where state welfare resources are overstretched and traditional models of family-based care are changing with the falling birth rate and the decline of multigenerational living.[5] A central finding is that the rise of visual digital communication among people in their fifties and older opens new possibilities for sociality and care among friends and family, with important implications, especially for women.

Drawing on 16 months of ethnographic research conducted primarily in a central urban site in Kyoto and a remote rural site in Kōchi

Prefecture, my research revealed how both sites share similar issues of depopulation. Indeed, studying a highly depopulated rural region offered a glimpse at the predicted future of Japanese cities. Living in depopulating areas where services are declining and there are fewer neighbours significantly impacts people's lives in both the rural and urban settings. However, it also engenders civic action and novel forms of community, as presented in Chapters 4 and 6. It was precisely because of depopulation that communities were designing interventions drawing on traditions of mutual care (*yui*). In this context, *Ageing with Smartphones in Japan* argues that the smartphone is becoming important in how people manage a sense of proximity and community while also pursuing independence. It is my hope that in doing so, this book extends recent anthropological work on ageing that centres values of interdependence in later life.[6]

My research suggests that a key mechanism through which people achieve a sense of proximity and communicate care at a distance is through visual digital media. In Chapters 4 and 5 I demonstrated how sociality in urban and rural settings can evoke feelings of intense scrutiny. However, people are aware that it is important to maintain and build social connections, especially as one reaches later life. The need to develop a social life was especially important for men after a lifetime of work. This finding was illustrated by Sakamoto-san, the man in his seventies who was a regular member of the dinner group in Kyoto. He observes:

> You can't just think about yourself when you are older. Meaning, you have to think of yourself in a group.

In this conclusion I draw together my arguments for how the smartphone is implicated in the ways that people practise forms of mutual support (*yui*) but at the same time employ what I call 'tactical invisibility' to participate in sociality while reducing the burden of that sociality through invisibility. I also extend my discussion of visual methods to approach digital technologies from the perspective of 'dwelling',[7] in order to frame how they are implicated in experiences of proximity at a distance. First, let me summarise the main ethnographic and theoretical contributions that I have aimed to provide in this book.

Ageing well

The topic of ageing offers rich scope for analysis when expanded to include the experiences of people in mid- to later life, i.e. those in their mid-forties to eighties, not the young or the 'old old'. In doing so, my research firstly questions assumptions about categories of age (Chapter 2) and looks at a continuum of experiences of ageing over the life course. As anthropologist Iza Kavedžija notes, if 'ageing well' means having a sense of wellbeing and meaningful relationships, is that so different to the desires of people at other times in their lives?[8] The people in this book did not define themselves by their ageing bodies, therefore 'ageing well' is shown to be more about feeling fulfilled and connected rather than something distinctly linked to their biological age. Nor did they deny physical decline and anxieties around care that can come with ageing, which I found to be increasingly mitigated through the smartphone (as shown in Chapters 2 and 5). I have endeavoured to provide the perspective of a diverse spectrum of individuals of different ages, showing how experiences of and attitudes towards ageing intersect with gender, socioeconomic status, family structures and personal aspirations.

Secondly, this book contributes towards an understanding of digital sociality in the lives of older people, foregrounding the notion of proximity at a distance via visual digital communication. The stories presented throughout show how negotiations of physical proximity are important to the lives of older people, fundamental to how they communicate (Chapter 4) and convey care through warm contact (Chapter 6). These in turn emerge as central to how meaning and purpose in life are manifested (Chapter 8). The smartphone is increasingly becoming enmeshed in these webs of care that connect people across physical spaces and across generations.

Thirdly, I have tried to show how experiences of ageing do not necessarily align with normative categorisations of ageing. In Chapter 2 I outlined the various ways in which people reject age-based labels or adapt them for their own purposes, challenging categorisation circulating in popular discourse and government policy such as 'pre-old age' (*zenki kōrei-sha*), 'old age' (*kōki kōrei-sha*) and 'super-elderly' (*chō kōrei-sha*).[9] Academic concepts of a 'third' and 'fourth' age[10] were demonstrated to be equally fluid. Similarly, 'successful ageing', rather than situated in a binary opposition with 'unsuccessful ageing',[11] was practised by individuals through their ability to care – both towards individuals and expressed in a wider sense through forms of labour (Chapters 7 and 8).

Ikigai, or purpose in life, was found to be just as subject to individual interpretation and contestation as other terms associated with ageing, as shown in Chapter 8. In a context in which a lack of *ikigai* in later life is a cause for public concern,[12] I argue that the term itself is contested by older adults and can be understood as standing in parallel, and thus sharing a similar critique, to that of 'successful ageing'[13] (Chapter 8). I argue that normative concepts such as *ikigai* and categories of age can be misleading; they are often highly contested in practice and are subject to change across the life course.

Normative roles for Japanese men and women, shaped by lifelong participation in highly gendered forms of labour, were similarly being challenged by people in later life when they faced the opportunities and challenges of retirement. The baby-boom generation, to which many of my participants belong, were the last generation to benefit from a corporate culture of lifelong employment and the economic boom years of Japan's bubble economy.[14] I found that many retirees, especially men, were reluctant to give up forms of labour completely, finding that their social networks and identities were bound up with work. Ideals of self-care and mutual care emerge as central to why many continued various forms of paid and unpaid labour, something that the anthropologist Shakuto Shiori argues is associated with neoliberal ideals of the recessionary era.[15] I complicate this picture by showing how a desire for sociality is at the heart of many people's continuation in various forms of labour for both men and women, as well a way to stay healthy in later life through having purpose (*ikigai*).

With policies aiming to achieve gender equality in the workplace having largely failed in Japan,[16] the smartphone is emerging as a site where women in particular can manage their various forms of paid and unpaid labour in later life. Despite many people having to follow mandatory retirement at the age of 65, for example public sector workers, many subsequently take on lower-paid, part-time positions or volunteer for community projects. I therefore found the term 'retiree' to be completely mismatched with the continued participation in the labour force that dominated so many of my participants' lives; it is appropriate only in referencing their ability to draw on their pension. Having said that, many anticipated future care costs and were eager to delay taking their pensions for as long as possible in order to receive a higher rate of pay.

This book has aimed to show how people in Japan are navigating sometimes oppressive societal expectations placed on older people, while finding individual forms of contentment and fulfilment in later life. Many people in this book had internalised some of the more negative public discourses regarding the burden that an ageing society presents. They would act in ways to mitigate the burden they themselves might impose or that they anticipated, including through the use of digital technologies. However, by their framing technology use as a way to lessen feelings of burden, is there a greater risk for new smartphone adopters, who are more vulnerable to the dangers of scams and fraud, to downplay negative encounters? If already highly attuned to reducing burden, might older adults be reluctant to ask for help when learning to use their device? In light of these remaining questions, there is a clear need for future research to approach older adults' technology use with sensitivity towards the wider social and cultural context within which they live, to ensure that their experiences, including any challenges or vulnerabilities, are fully acknowledged and addressed.

My research was primarily with people around middle-age who tended to be caring for older family members, often at a distance. They were not immune to negative attitudes towards ageing, reflected in their fears about getting old and a desire not to live 'too long'. Yet this book has demonstrated the many positive things that can come with ageing, and indeed from the imperative to care. While this book has not focused on the care of the 'old old', it has reflected more broadly on how an attitude of care infuses and enriches later life. In caring for one's community, or for one's friends, even if that is simply through the exchange of LINE messages, people make their lives meaningful.

In the past 15 years there has been renewed scholarly attention on care in fields both within and outside those typically associated with the topic, such as in science and technology studies (STS).[17] The multidisciplinary Care Collective, authors of *The Care Manifesto*, emphasise the need for a politics centred on care in response to the world's carelessness, particularly as highlighted by the Covid-19 pandemic. They argue that we have neglected caring for each other, especially the vulnerable, poor and weak. In focusing on innovations by communities navigating challenging demographic issues, I have shown how people are reclaiming collective and communal life to foster what the Care Collective call a 'radical cosmopolitan conviviality',[18] while balancing personal autonomy through the digital. The emerging centrality of personal digital devices within the practice of care at a distance and mutual support offers a rich field for further study.

Proximity at a distance

Changing domestic arrangements that favour single living and the depopulation of both rural towns and urban centres, including Kyoto, are seen by many to be contributing to the weakening of 'horizontal relationships' (*yoko no kankei*). This concern can be understood more broadly as related to fears over Japan's 'relationless society' (*muen shakai*).[19] I found that alongside such changes were shifting attitudes towards traditional roles for men and women during the life course which were being impacted by, and reflected in, smartphone use. I found that physical proximity tended to be central to the sociality of men (Chapter 4), whereas women tended to be more embracing of the affordances of the smartphone for communicating friendship and connection at a distance (Chapter 5). However, for men and women alike, a nostalgia for community rooted in physical proximity fuelled civic action to reconstitute *yui*, or mutual relations, often in novel ways. While the concept of *yui* has agricultural roots, I demonstrated in Chapter 5 how it is still important in contemporary Japan, represented in organised forms of sharing and co-operation.[20]

In this context, the smartphone, while being a device that for many represented connection, was often simultaneously framed as part of this shift towards individual living, precisely because of the way that it enabled remote connection. However, in observing how people navigated relationships of care, both within multigenerational households (Chapter 2) and remotely between kin and non-kin (Chapter 5), a similarity emerged in the way that distance was inserted into relationships of care. Building on anthropological work on the importance of distance and space in relationships among older adults in Japan,[21] and wider work on the centrality of 'wrapping' in Japanese sociality,[22] *Ageing with Smartphones in Japan* extends discussions around the importance of form in visual digital communication. More precisely, I argue that it is through the visual digital medium of stickers that the smartphone becomes an affective object through the way in which they create 'atmosphere' – akin to offline atmospheres interpreted through bodily and facial cues that rely on physical proximity. Stickers and emojis, while communicative of emotion, are also a form of facade. They serve to generate a space between people that facilitates a sense of ambivalence, something that is central to face-to-face communication in Japan.

Anthropologists Ohashi, Kato and Hjorth have demonstrated how messaging maintains matriarchal intergenerational relationships across geographical distances, arguing that 'LINE is *the* form of intergenerational communication for Japanese by enveloping mundane intimacies with hybrid forms of new media literacy'.[23] Building on previous work that situated mobile phones, and messaging in particular, as mediators of 'ambient virtual co-presence',[24] Ohashi et al. argue that emojis and stickers facilitate care at a distance, constituting a form of digital kinship. I have extended the focus on the affordances of stickers to show how they are particularly useful among women in later life as a medium of care inside and outside of the family. It is in forms of mutual support through all-female friendship groups that older women use stickers to display personality and affirm their connection to each other via visual means (Chapter 5). Such relationships are considered so important that they are often what kept many older women in precarious forms of labour, as I argued in Chapter 3.

Returning to the concept of *yui*, which traditionally relied on physical proximity, I argue that stickers are a medium through which people tie themselves together across distances in mutually supportive relationships. The emotion work[25] that stickers afford makes them particularly useful in relationships of care, such as between elderly parents and their middle-aged children, as seen in Chapter 5, to facilitate more frequent yet less burdensome contact. Just as people carefully manage their relationships of care within a domestic space,[26] social proximity via LINE is affectively managed through stickers and emojis that convey affect, but reduce the emotional cost of over-closeness.

Some people, such as the woman who loved watching figure skating videos in Chapter 5, even employed the smartphone to make themselves invisible to those closest to them, in what I term 'tactical invisibility' – a form of privacy only possible because of their lack of physical proximity. However, the 'trap' of visibility[27] that the smartphone enables can also be the very mechanism through which the smartphone can become a positive tool for intervention, as discussed in Chapter 6 in relation to health.

Just as rural Japan is being recognised for its diversity, permeability and vitality as a location of 'heterotopia',[28] in which migrants and international networks flow, the smartphone can similarly be understood as facilitating a diversification of digital spaces through its rapid uptake by people of all ages. Older smartphone adopters, often bypassing computers altogether, are forging their own digital cultures. Their movement online parallels the intentions of urban-to-rural migrants, as discussed in Chapter 7, in the search for both connection and autonomy. This book has demonstrated how the Japanese countryside in rural Kōchi Prefecture is a highly dynamic place, with flows of people and ideas fuelling civic action around rural revitalisation efforts. Migrants originating in cities ('I-turners') and those returning from working in cities ('U-turners') have featured in many recent ethnographies of rural Japan;[29] they are attracted by the low costs of living associated with rural depopulation. People make the move not only for financial reasons, but also, as I have shown, to find a sense of belonging and security within a wider context of precarity.

Being physically close to a supportive network of people is important, especially as a person ages. 'The internet is not your neighbourhood,' declared Hayashi-san in Chapter 4, yet I have tried to show how it can be integral to how people craft spaces of belonging, both online and offline. My research suggests that digital communication via the smartphone is shifting notions of what home and proximity mean in later life.[30] This shift can be seen in private homes, as with Takeshi-san, and in semi-public spaces, such as Kazuko-san's home and event space in Kyoto (Chapter 4), both of which relied on digital means to bring more sociality into the domestic sphere. As introduced in Chapter 2, changing living arrangements have meant that previously large, multigenerational homes are now subject to new uses (for example, via platforms such as Airbnb) and meanings, often in conjunction with digital technologies that can enable people to reconstitute some of the sociality they fear is lost with such shifts.

Drawing together these findings on proximity and digital sociality, I turn to the work of anthropologist Tim Ingold, who invites us to ponder what it means 'to dwell'.[31] Influenced by the writings of Heiddegger and Merleau-Ponty, Ingold contends that 'built' environments are constituted through the daily life activities of the people who dwell within them, through which 'the world continually comes into being around the inhabitant'.[32] According to Ingold, a dwelling perspective reveals how the act of building is processual, with architecture such as houses regarded as evolving forms. Following this line of thought, I argue that it is through inhabiting personal devices such as the smartphone, that digital space becomes meaningful for people.

One of the means by which we digitally 'dwell' is through the affordances of visual digital communication that enable us to actively shape our digital worlds. By sharing visual media, individuals actively build their digital landscapes in ways that are personal and affective, creating 'atmospheres' that recall offline embodied social interactions. In doing so, I follow Ingold in questioning the distinction between 'human' and 'artificial' that so troubles contemporary popular discourse with the rise of artificial intelligence (AI). I argue that technological artefacts such as smartphones are not static entities but evolve alongside human engagement, ultimately challenging traditional notions of technological determinism and highlighting the dynamic, ongoing nature of human-technology relationships.

A central finding from the multi-sited wider project of which this research is part was that the smartphone is invisibly anthropomorphic; the device is shaped through its usage, but it also shapes the user, becoming both an extension of a person and also their reflection.[33] I argue that by also paying attention to non-usage we can learn how a rejection of the device and its affordances reflects, and also shapes, a person's social world. In this light I contend that gendered differences, particularly in the usage or rejection of digital visual media, as outlined in Chapter 5, can mean that men are less able to participate in this mechanism for digital dwelling, resulting in a wider rejection of the smartphone.

The older men in my research tended to not feel at home in the smartphone, paralleling the contentious relationship many had to their physical home. Women on the other hand tended to embrace the affective capacities of stickers and emojis as a means to create a comfortable dwelling space in which social relationships could unfold more akin to in offline settings. I argue that it is practices of care that constitute the smartphone as a place of dwelling, much as within the physical home. In order to investigate the embodied ways that we dwell in and with digital devices I propose a graphic methodology, to which I now turn.

A graphic approach

In response to the shift in sociality towards the visual that corresponds with the uptake of digital devices, I have developed a range of graphic ethnographic approaches. These include participatory drawing and methods that exploit the visual affordances of smartphones themselves. The prevalence of illustrated media in contemporary life in Japan, including manga and now the daily exchange of billions of LINE stickers, means that as an anthropologist I needed to find a similarly visual means to explore the smartphone. How does one 'follow the smartphone' or indeed 'follow the sticker'?[34]

Smartphones were most used by my interlocutors in private moments, away from my observation. Through conducting 25 smartphone interviews with people, I learned about the apps that they use and their feelings towards the device. However, in order to develop a method that would enable me to understand how the smartphone moves with a person throughout the day, I needed to turn to visual affordances of the smartphone itself – in this case the screenshot.

An important affordance of the platform LINE are features that facilitate visual communication. Through LINE video calls I was able to continue my research remotely from the UK when I returned from fieldwork in 2019. This communication continued and took on new significance when the global Covid-19 pandemic emerged in 2020, confining my research participants and myself to our homes. Through regular digital interaction we supported each other through those strange times. While I was no longer in Japan, it was during this time of limited physical mobility that my adapted digital research methods enabled me to learn things that I did not when I was physically in the country. The constraints of the pandemic on researchers have spawned 'agile methods',[35] with some researchers turning to artistic collaboration as a means to study sensory, affective and relational phenomena at a distance.[36] In this manner I invited several research participants to collaborate on the short videos that have appeared throughout this book.

Over the past 10 years I have been inspired by the proliferation of graphic ethnographic research approaches in anthropology and related disciplines.[37] Anthropologist Dimitrios Theodossopoulos offers a comprehensive overview of the emergence of graphic ethnography, suggesting that, with its myriad affordances, it merges descriptive, applied and analytical approaches, challenging authorial authority and power dynamics, thereby expanding creative and analytical freedom in ethnographic representation.[38] After the pandemic had receded somewhat and things were starting to return to normal in Japan, I invited Maiko-san, a woman in her sixties living in Kyoto, introduced in Chapter 4, to collaborate on a graphic story about her new smartphone. In the comic at the start of this chapter we see how her smartphone use is situated within a wider ecology of devices and relationships.

To produce the comic, Maiko-san kept a visual diary of her smartphone use for one day through taking screenshots each time she used her phone or sending me photos that she took. This resulted in 16 images which we then discussed together at length over LINE video calls and messages. I asked her questions about where she was when she took each screenshot, how she was feeling and what her motives behind her usage were, etc., in order to gain a comprehensive understanding of her digital interactions over 24 hours. I wanted to understand how digital repertoires develop across devices and are situated within physical and embodied contexts.

In the comic we see how Maiko-san uses her laptop and smartphone simultaneously to send and reply to textual messages, send a photo and browse websites. Through this graphic form we also learn a range of other insights about Maiko-san's technology use, such as how she keeps her phone and laptop charging together overnight, the fact that she does her morning exercises while waiting for her laptop to start up, and her enjoyment of the warmth of the laptop on her lap as she sits in her armchair to scroll through the morning news. We see how she interchangeably sends messages via speech-to-text through her smartphone, replies to messages using the web interface of her messaging application and then sends a photo directly from her phone. In creating the comic with Maiko-san I discovered that her configuration of devices is as bound up with their limitations as their affordances. Her laptop takes a long time to start up, which is the reason why her phone is the first device she uses while she waits. Her phone's keyboard is too small and difficult to type on, which is why she prefers using her laptop to send messages. Through the comic we can see how digital repertoires are developed in embodied and emplaced ways, focusing attention on the material dimension of technology use.[39]

Maiko-san's physical proximity to her multiple devices, or 'screen ecology', and the way in which she engages with them in the space of her home are visualised; so is the audio notification when she receives a message. In this way, the graphic form enables a multisensory presentation of her morning routine with her devices in a manner which allows the viewer to dwell in each moment and absorb details, making connections for themselves between panels.[40] In making the comic, I was able to present a draft version back to Maiko-san. Her suggestions were then incorporated into the final draft. Presenting a visual narrative back to Maiko-san in a form that she could quickly engage with enhanced the collaborative and participatory nature of this exercise. It also provided further insights regarding her thoughts and feelings which had not come through in a prior interview with her about her smartphone.

Taking inspiration from the proliferation of visual communication among my interlocutors, the production of a graphic narrative as a form of multimodal collaboration was a logical step, meeting them via a medium with which they were already familiar. Prior to getting a smartphone, Maiko-san had used a feature phone (*garakei*), with which she would often take photos when we went on walks around Kyoto. One day when walking around a temple, she pointed to a window that perfectly framed the garden behind it, commenting 'Japanese people love to frame things; it's in our culture'. The framing of a garden performed by the architecture of a temple is similar to the way in which digital photography enables people to frame their experiences. Even if they never look at the photo again, as Maiko-san said of the photo she took of her flower arrangement, featured in the comic, it is the act of framing itself that directs attention in the moment. With the panels of comics serving as 'windows of attention',[41] framing the action or scene, we can see how the rise of digital photography, and thus familiarity with framing in general, may make graphic narratives more accessible even to those not previously accustomed to the genre.

With the emergence of smartphones and other digital devices as ubiquitous across many populations, research participants often share the same skills and tools of representation as anthropologists.[42] Even when skills may not be comparable, the resulting difficulties or discrepancies can also be opportunities for insight, as in the production of the comic described above. Initially Maiko-san did not know how to take a screenshot, and I was unfamiliar with her 'simple' smartphone android device. Her immediate response was to search on Google for how to take a screenshot, which led to a discussion of how she often watches educational YouTube videos when she does not know how to do something.

She is self-reliant, yet the comic reveals how Maiko-san is embedded in a network of relationships of care. There is care between the *ikebana* teacher and Maiko-san, care from commercial sources (the bento delivery service) and care that Maiko-san is showing to her friends by sending them messages, songs and photos. We see how care is embodied in the way that Maiko-san and her teacher bow to each other when Maiko-san departs, which is then mirrored in the body language of stickers. There is also the care that Maiko-san is performing for the sake of her health. This is aided by checking her step counter, which motivates her to walk more the next day. The smartphone enables Maiko-san to be embedded in those practices of care which cannot be classified as 'online' or 'offline', but are instead shown here to be located physically and digitally simultaneously. We see how, despite living alone, Maiko-san is close to others, with affective embodied gestures communicated through visual messaging.

In developing a graphic approach to studying Maiko-san's use of her smartphone throughout one day, I am drawing upon scholarship in visual anthropology that urges a diversification from the centrality of text.[43] Graphic approaches can be a key offering in multimodal anthropology, given the way in which it is possible to convey multisensorial and embodied experiences in a medium which can be easily reproduced on the page, but which is not limited to text. In the making of the comic with Maiko-san I was able to 'witness' the private moments in which screen ecologies really came into play.

The graphic form allows a viewer to see the kinds of visual communication in which Maiko-san is engaged in an immersive format. Multimodal research methods 'generate relations' and foster forms of collaboration that can make for a 'more public, more collaborative, more political' anthropology.[44] I would argue that such forms also make it more possible for empathetic engagement of audiences with our research subjects. Mundane experiences with technologies often occur in moments when a person is alone, making them difficult for a researcher to access.[45] I found that using the visual affordances of the smartphone, more precisely the screenshot diary, as an elicitation device enabled me to situate Maiko-san's embodied digital repertoires in space and time.[46] This method also allowed her to be able to frame what she wanted to share with me as a researcher. A screenshot diary put control into the hands of the participant, enabling her to share what she did with her multiple devices without challenging her privacy.

Similarly, in asking Megumi-san to create a painting to respond to the term *ikigai* in Chapter 8 I was inviting her to reflect on and direct the subsequent discussion according to her understanding of the topic and the stories she wanted to tell. Accordingly, my research is part of a wider movement of graphic research methodologies that allow for some reflective distance between the research participant and potentially sensitive topics under discussion, which paradoxically can enable a deeper discussion.[47] Megumi-san could spend time deciding on the topics she wanted to talk about through her narrative painting, which she then led the discussion on.

In the other instances of participatory drawing shown in Chapter 5, participants were similarly encouraged to reflect on the topic that was to be under discussion through visual means.[48] Drawing allowed participants to tap into feelings that may be difficult to articulate in words, such as their experiences of *honne* and *tatemae* and their relationship to the smartphone, adding an important corporeal dimension to the discussions that followed.

This book is replete with examples of how older Japanese people are themselves developing their uses of the smartphone, including forms of visual digital communication, to enable continuity with certain aspects of idealised sociality referenced as *yui*. By incorporating visual communication as a form of distancing and engaging in practices of 'tactical invisibility', they can extend forms of sociality thought to be under threat while reducing over-closeness. However, the usage of visual digital communication in the form of stickers commodifies affective labour.[49] Stark and Crawford suggest that stickers and emojis are 'more than just a cute way of "humanizing" the platforms we inhabit: they also remind us of how informational capital continually seeks to instrumentalize, analyze, monetize, and standardize affect'.[50]

As such, participation in visual digital communication is different from the kinds of drawings produced during the elicitation exercises described above. However, upon visually analysing the drawings produced, in particular the drawing in Chapter 5 (Fig. 5.6) showing the range of emotions experienced via the smartphone, it is clear to see how styles of drawing are shaped by wider visual consumption habits. The influence of manga, and even of stickers, in how this particular participant represented her own range of emotions speaks to the importance of acknowledging how the expression of emotional repertoires can be shaped by digital aesthetic repertoires, themselves embedded in histories of visual culture and consumption.

Multimodal approaches to the study of human-technological entanglements demonstrate the value of co-created material that centres the experiences of research participants. As such, this book contributes to the growing recognition of collaboration in anthropological research, not only with research participants, but also across disciplinary divides. Heffernan et al. state that such collaboration is crucial to anthropology as a discipline if it is to remain relevant and sustainable.[51] Graphic approaches can facilitate a more collaborative research practice that embeds participants' own visual representations in analysis and dissemination. This drive towards collaboration was also behind the mHealth intervention mentioned in Chapter 6, in which the ethnography and practical intervention informed one another, rather than being situated at opposite ends of anthropological practice. With the rise of smartphones and access to digital devices now ubiquitous across most parts of the world, anthropologists are presented with a vast opportunity for collaborative engagement, both during on-site and remote fieldwork.

The illustrations that appear at the start of each chapter of this book were made during the process of editing this book. I used a combination of sketches from fieldwork and imaginative drawing to produce images that I hope contribute to the narrative flow of the book, albeit while being somewhat ambiguous, without corresponding captions. In a similar way to fieldwork, in which experiences and observations can have many meanings, I invite the reader to come up with their own interpretations of the illustrations. The images are intended as a moment to dwell with the people who populate these pages and who, apart from those who appear in the videos, remain invisible and anonymous.

In their creation, the drawings also served an analytical purpose. While drawing, I found that I made connections in my research material that were bubbling under the surface but which, thus far, had not emerged in my writing. For example, the image that appears at the start of Chapter 4 illustrates the feeling of close scrutiny that many people shared. This concern for what other people think does not lessen with age. As Akiko-san observed, 'the eyes that surround you will be around you forever'. The red lines in the image were intended to emphasise the claustrophobic sensation of being surrounded and watched.

When it came to illustrating the concept of *ikigai* in Chapter 8, I started with a drawing by 70-year-old Yamada-san who sketched *ikigai* as a triangular web of smiling faces – being connected to other people and making them smile was what gave her a sense of purpose in life. Inspired by her visual idea, I played with the triangular web until I realised it looked like a mountain. I then realised what she was talking about was similar to the concept of *yui*, or reciprocal care, with its roots in rice farming. The red line in the resulting illustration indicates the reciprocal relationships between farmers that binds rural communities together. The red lines in both images thus represent care as much as scrutiny (Fig. 9.1). In thinking about these concepts visually I came to understand that *yui* and the intense feeling of being observed are two sides of the same coin – one cannot exist without the other. It is this orientation towards others that can make people uneasy, but which also facilitates connection and is crucial to how people care for themselves and each other throughout their lives.

Figure 9.1   Illustrations of close social scrutiny (on the left) and of *ikigai/yui* (on the right). Images by Laura Haapio-Kirk.

Conclusion

In this book I have endeavoured to shed light on the intersection of digital technology, ageing and the evolving landscape of care in contemporary Japan. The ethnographic fieldwork that informs this book explored the daily lives of people in mid-life and above, showing how they face the challenges of an ageing population and increasing demands for care. In doing so this research highlights the importance of mutual support networks in navigating the dual crisis of depopulation and ageing which face not only Japan, but much of the world. Being at the forefront of these demographic changes, Japan can offer a window onto not only the implications of these challenges, but also how people creatively navigate them, increasingly with the aid of their smartphones.

As highlighted throughout this book, the digital sphere offers avenues for connection, resource sharing and emotional support – essential components in addressing the multifaceted needs of an ageing population. In spending 16 months with older people as they adopted or rejected the smartphone, I also saw the flipside of the device. People were highly concerned about the potential negative implications of digital communication for their relationships and for how they were perceived as social persons. This was often linked to a wider concern around presenting themselves as 'good' ageing persons by staying healthy and caring for others. I witnessed how both the so-called 'young-old', those who were active and productive, and 'old-old', those who were more dependent on care, were equally subject to forms of societal othering in the ways in which certain traits or attitudes, such as an orientation towards *ikigai* (life purpose), were glorified in later life.[52] It is the responsibility of anthropologists, whose expertise is precisely to dismantle such things as moral expectations around ageing and indeed age categories themselves, to reveal alternative stories about how people find their own ways of living and ageing well despite challenges.

This book began with an account of the first day of the new *Reiwa* era in rural Tosa-chō. The people I spent time with on that day were a mix of urban migrants ('I-turners'), return migrants ('U-turners') and locals who had lived the majority of their lives in the region. When thinking about this mix of people and the diversity of their lives, it is difficult to make broader comparative generalisations regarding the way in which later life is experienced in cities compared with the countryside. In many ways, the blurring of the urban-rural boundary mirrors the way that the life course itself is experienced: as fluid and with no distinct boundaries between young and old. The same issues that face older adults in urban and rural areas, such as loneliness and isolation, also exist for younger people. In the countryside communities are idealised for their commitment to *yui* and simultaneously felt to be over-close. Depopulation in both rural and urban Japan creates challenges to physically proximite care. Yet it was precisely the community spirit galvanised by such issues that was drawing migrants in search of connection from cities to the countryside.

In a country with frequent natural disasters, the smartphone was adopted by some older people as a safety net, particularly for those living alone, showing how it can make people feel more secure. But what wider societal failings does this point towards, and potentially enable? In Japan, where welfare systems are strained and societal shifts are reshaping familial dynamics, digital technologies such as the smartphone are emerging as important for the communication of care. However, people were at the same time concerned that they might facilitate a withdrawal of face-to-face connection.

This book has shown how older people are adopting the smartphone as givers of care, rather than being only recipients; they are also active in community building, rather than merely depending on the community. As anthropologist Jason Danely suggests, it is important to nuance the notion of 'care givers and receivers', as this does not capture 'the fluidity and ambiguity of the intimacy and responsiveness within caring relationships'.[53] The evolving role of smartphones in facilitating care underscores the need for a more comprehensive understanding of caregiving dynamics in the digital age.

The Covid-19 pandemic has highlighted the urgency of reimagining care paradigms and embracing innovative solutions to support older adults in an era of social distancing and heightened vulnerability. Within this context, the importance of mutual support networks facilitated by smartphones becomes increasingly evident. The pandemic revealed to people around the world the possibilities for connection afforded by the digital at a time when physical proximity was impossible. Despite the relatively low prevalence of the virus in Japan, especially in rural areas, the social pressure to show self-restraint (*jishuku*) by staying at home, as encouraged in public state discourse, was immense. The Social Welfare Office in Tosa-chō began LINE messaging those who they would otherwise visit at home. Several people told me how they had taken up video calling with greater frequency because of their reluctance to visit one another. The smartphone, rather than representing an 'inhuman' mediator with 'no heart', as several people had lamented prior to the pandemic, became a much-relied upon conduit of feeling and affection.

My on-site fieldwork ended just before the Covid-19 pandemic began. Since then, people of all ages around the world have faced the kinds of issues previously associated mainly with older adults: those of isolation, reduced mobility and vulnerability. We have all come to realise more directly the importance of interdependence and mutual support that lies at the heart of care. Medical anthropologist Arthur Kleinman has argued that it is care giving and receiving that makes us human,[54] and so perhaps the pandemic, as with other traumatic events such as the earthquake, tsunami and nuclear meltdown that occurred on 3 March 2011 (3.11) in Japan, will offer a catalyst for us to become more relational, more caring and thus more human.

A worker at the town hall in Tosa-chō explained that during the pandemic he had helped several local businesses to go online. One couple who grew flowers and would usually sell them locally were at first hesitant to sell their produce online because they valued the interaction that they usually had when dealing with customers face to face. However, their minds were changed when customers as far as Osaka and Tokyo would call and write to them to express their delight at the flowers they had received. The town hall worker explained:

> Before they were kind of allergic to the internet, but after we did this together, they want to keep doing it. They said they thought the internet was more dry and anti-human like. They got many letters and postcards and emails to thank them and encourage them in this corona time. That changed their minds about the internet. Not dry, not anti-humanity.

It has been argued that the triple disaster of 3.11 is directly linked to the expansion of mobile media use for co-presence and care in Japan.[55] The long-term implications of the pandemic on the trajectory of digital devices are yet to be seen, but my research before and during the pandemic reveals that older people in Japan were already engaged in digital forms of care and affectivity, and this only increased during the crisis.[56] The issue of depopulation is similarly challenging, especially in rural regions, yet my aim in this book was to show how communities are creatively responding and even thriving in such contexts. While at the time of my fieldwork the smartphone was only beginning to be taken up by older people, it was clear that its promise of connection would become even more cherished with increasing depopulation.

The rise of smartphones among all demographics has profound implications, not only for interpersonal relationships but for how we are engaged in constituting ourselves. This book attempts to join other anthropological work on ageing populations around the world in reframing the negative narratives of ageing by highlighting the opportunities for self-development that can come with living a long time.[57] The people in *Ageing with Smartphones in Japan* variously demonstrate how later life can grant more time for relationships outside of the confines of work or family life, and can be a period of reinvention and flourishing. While acknowledging the challenging societal context in which many people are ageing and experiencing individual age-based issues, this book has aimed to show how later life contains possibilities for living well, rather than 'ageing well', centred on practices of care that are increasingly grounded in the digital.

Notes

1. A pseudonym.
2. Daniels 2024.
3. Danely 2019; Kavedžija 2019.
4. Dobashi 2005; Amagasa 2012 (as cited in Ohashi et al. 2017; Matsuda 2009; Ohashi et al. 2017.
5. Suzuki 2015.
6. Kavedžija 2019; Lamb 2014.
7. Ingold 2000.
8. Kavedžija 2019.
9. Ouchi et al. 2017.
10. Gilleard and Higgs 2010.
11. Rubinstein and de Medeiros 2015.
12. Kavedžija 2019.
13. Rubinstein and de Medeiros 2015.
14. Hirayama and Ronald 2008.
15. Shakuto 2018.
16. Crawford 2021.
17. Lindén and Lydahl 2021.
18. The Care Collective 2020, 20.
19. Allison 2017.
20. Suehara 2006.
21. Kavedžija 2018.
22. Hendry 1993.
23. Ohashi et al. 2017, 1 (original emphasis).
24. Ito and Okabe 2005.
25. Hochschild 1979.
26. Daniels 2015.
27. Foucault 1977.
28. Hansen and Klien 2022.
29. Rosenberger 2017; Klien 2020; Manzenreiter et al. 2020; Traphagan 2020.
30. A key finding across fieldsites from the wider ASSA project, of which my project is part, is that the smartphone has become a 'transportal home' where people can dwell when not actually in their physical homes. This leads in turn to a 'death of proximity', in which people might be spending time with others through their phones rather than the people they are sitting next to (Miller et al. 2021, 219–27).
31. Ingold 2000.

32. Ingold 2000, 153.
33. Read more about the theoretical findings from the wider ASSA project, including 'beyond anthropomorphism' and the 'transportal home' in the concluding chapter of *The Global Smartphone* (Miller et al. 2021).
34. Marcus 1995.
35. Watson and Lupton 2022.
36. Bates et al. 2023.
37. In 2020 I co-curated the exhibition *Illustrating Anthropology* (https://illustratinganth ropologycom.wordpress.com/) with Dr Jennifer Cearns and with the support of the Royal Anthropological Institute of Great Britain and Ireland, where we were Leach Fellows in Public Anthropology. The exhibition has received over 20,000 visitors and has generated a lively Instagram community. It explores human lives around the world through comics, drawings and paintings of anthropological research. From those who use illustration as a fieldwork method to others who partner with artists and research participants to tell stories, the exhibition draws together a wide range of ways in which contemporary anthropologists are illustrating anthropology.
38. Theodossopoulos 2024.
39. Digital anthropology emerges directly from the field of material culture studies, in which objects are not viewed merely as accessories to human existence but are also recognised as pivotal in mediating our humanity. Digital technologies are thus seen as no less or more mediating than other kinds of material culture (Miller and Horst 2020).
40. McCloud 1993. As comics theorist McCloud argues, it is in this imaginative work that happens between the panels that the power of comics lies.
41. Cohn 2014.
42. Favero and Theunissen 2018.
43. Banks and Zeitlyn 2015; Cox et al. 2016; Pink 2001, 2009; Stoller 2010.
44. Dattatreyan and Marrero-Guillamón 2019, 221.
45. Hjorth et al. 2020.
46. This use of visual collaboration as a tool of elicitation was based on my observation of participants' visual digital practices. Likewise the design of the health intervention, discussed in Chapter 6, was based on my ethnographic understanding of wider modes of visual digital sociality and of food practices. This ethnographic grounding of the intervention in existing practices was probably instrumental in its organic continuation beyond the proposed intervention period.
47. Some other examples include making use of metaphor (Warburton 2005), satire (Moloney et al. 2013) and stereotypes (Macgillivray 2005) to approach sensitive research topics.
48. Gauntlett 2007.
49. Stark and Crawford 2015.
50. Stark and Crawford 2015, 8.
51. Heffernan et al. 2020.
52. Van Dyk 2016.
53. Danely 2023, 13.
54. Kleinman 2009.
55. Ohashi et al. 2017; Hjorth and Kim 2011.
56. I have published elsewhere graphic representations of the results of my remote research on the impact of the Covid-19 pandemic on smartphone use in Japan (Haapio-Kirk 2020).
57. Kavedžija 2019.

References

Ahlin, T. 2020. 'Frequent callers: "Good care" with ICTs in Indian Transnational Families', *Medical Anthropology* 39(1): 69–82. https://doi.org/10.1080/01459740.2018.1532424.

Ahmed, S. 2010. *The Promise of Happiness*. Durham, NC: Duke University Press.

Ajana, B. 2017. 'Digital health and the biopolitics of the quantified self', *Digital Health* 3: 1–18.

Akimoto, A. 2013. 'Looking at 2013's Japanese social-media scene', *The Japan Times*, 17 December 2013. https://www.japantimes.co.jp/life/2013/12/17/digital/looking-at-2013s-japanese-social-media-scene-3/.

Aldrich, D. P. 2011. 'The power of people: Social capital's role in recovery from the 1995 Kobe earthquake', *Natural Hazards* 56(3): 595–611. https://doi.org/ 10.1007/s11069-010-9577-7.

Alexy, A. 2011. 'Intimate dependence and its risks in neoliberal Japan', *Anthropological Quarterly* 84(4): 895–917.

Allison, A. 2013. *Precarious Japan*. Durham, NC: Duke University Press.

Allison, A. 2017. 'Greeting the dead: Managing solitary existence in Japan', *Social Text* 35(1): 17–35. https://doi.org/10.1215/01642472-3727972.

Amagasa, K. 2012. 'Keitai to Kazoku' ['Keitai and family']. In *Keitai shakairon* [Keitai society], edited by T. Okada and M. Matsuda, 101–116. Tokyo, Japan: Yuhikaku.

Amrith, M., V. K. Sakti and D. Sampaio. 2023. 'Introduction'. In *Aspiring in Later Life: Movements Across Time, Space, and Generations*, edited by Megha Amrith, Victoria K. Sakti and Dora Sampaio (1st ed.), 1–18. New Brunswick, NJ: Rutgers University Press. https://doi.org/10.36019/9781978830431.

Ardèvol, E. and E. Gómez-Cruz. 2013. 'Digital ethnography and media practices'. In *The International Encyclopedia of Media Studies*, edited by A. N. Valdivia, 498–518. Hoboken, NJ: John Wiley. https://doi.org/10.1002/9781444361506.wbiems193.

Ariga, K. 1957. 'Yui no imi to sono henka' ('The meaning of Yui and its transformation'), *The Japanese Journal of Ethnology* 21(4).

Aronsson, A. 2022. 'Professional women and elder care in contemporary Japan: Anxiety and the move toward technocare', *Anthropology & Aging* 43(1): 17–34.

Asao, Y. 2007. 'Outlook on the retirement process of *dankai no sedai*, or the Japanese baby-boom generation', *Japan Labor Review* 4(4): 121.

Atsumi, T. and J. D. Goltz. 2014. 'Fifteen years of disaster volunteers in Japan: A longitudinal fieldwork assessment of a disaster non-profit organization', *International Journal of Mass Emergencies & Disasters* 32(1): 220–40.

Azevedo, A. and M. J. Ramos. 2016. 'Drawing close. On visual engagements in fieldwork, drawing workshops and the anthropological imagination', *Visual Ethnography, Participatory Approaches to Visual Ethnography from the Digital to the Handmade*. https://doi.org/10.12835/ve2016.1-0061.

Azuma, K. 1999. *Eijingu no Shinrigaku* [*Psychology of Ageing*]. Tokyo: Waseda University Press.

Babu, S. S., S. Mishra and B. B. Parida. 2008. *Tourism Development Revisited: Concepts, issues and paradigms*. Delhi: Sage Publications India.

Bachnik, J. M. 1992. 'The two "faces" of self and society in Japan', *Ethos* 20(1): 3–32. https://doi.org/10.1525/eth.1992.20.1.02a00010.

Backhaus, P. 2008. 'Chapter Twenty-Three. Coming to terms with age: Some linguistic consequences of population ageing'. In *The Demographic Challenge: A handbook about Japan*, edited by F. Coulmas, H. Conrad, A. Schad-Seifert and G. Vogt, 455–71. Leiden: Brill.

Ballantyne, A., L. Trenwith, S. Zubrinich and M. Corlis. 2010. '"I feel less lonely": What older people say about participating in a social networking website', *Quality in Ageing and Older Adults* 11(3): 25–35. https://doi.org/10.5042/qiaoa.2010.0526.

Banks, M. and D. Zeitlyn. 2015. *Visual Methods in Social Research* (2nd ed.). London: Sage. https://doi.org/10.4135/9781473921702.

Bartlett, R. and A. Kafer. 2020. 'Thinking into aging–disability nexuses: A dialogue between two scholars'. In *The Aging–Disability Nexus*, edited by K. Aubrecht, C. Kelly and C. Rice, 251–67. Vancouver, BC: UBC Press.

Baseel, C. 2019. 'Japan announces new era name, Reiwa, but what does it mean and why was it chosen?', *Sora News* 24. 1 April 2019. https://soranews24.com/2019/04/01/japan-announces-new-era-name-reiwa-but-what-does-it-mean-and-why-was-it-chosen/.

Bass, S. A. 1996. 'An overview of work, retirement, and pensions in Japan', *Journal of Aging & Social Policy* 8:2–3, 57–78. https://doi.org/10.1300/J031v08n02_05.

Bates, C., K. Moles and L. M. Kroese. 2023. 'Animating sociology'. *The Sociological Review* 71(5), 976–91. https://doi.org/10.1177/00380261231156688.

Befu, H. 1980. 'A critique of the group model of Japanese society', *Social Analysis: The international journal of social and cultural practice* 5/6: 29–43.

Befu, H. 1993. 'Nationalism and nihonjinron'. In *Cultural Nationalism in East Asia: Representation and identity*, edited by Harumi Befu, 107–35. Berkeley, CA: Institute of East Asian Studies, University of California.

Befu, H. 2002. 'Symbols of nationalism and *nihonjinron*'. In *Ideology and Practice in Modern Japan*, edited by R. Goodman and K. Refsing, 40–60. Abingdon, Oxon; New York, NY: Routledge.

Bell, C., C, Fausset, S. Farmer, J. Nguyen, L. Harley and W. B. Fain. 2013. 'Examining social media use among older adults', *Proceedings of the 24th ACM Conference on Hypertext and Social Media*: 158–63. https://doi.org/10.1145/2481492.2481509.

Ben-Ari, E. 2013. *Changing Japanese Suburbia*. Abingdon, Oxon; New York, NY: Routledge.

Benedict, R. 2005. *The Chrysanthemum and the Sword: Patterns of Japanese culture*. Boston, MA: Houghton Mifflin Harcourt.

Benson, M. 2013. *The British in Rural France: Lifestyle migration and the ongoing quest for a better way of life*. Manchester: Manchester University Press.

Benson, M. and K. O'Reilly. 2009. 'Migration and the search for a better way of life: A critical exploration of lifestyle migration', *The Sociological Review* 57(4): 608–625. https://doi.org/10.1111/j.1467-954X.2009.01864.x.

Berlant, L. 2011. *Cruel Optimism*. Durham, NC: Duke University Press. https://doi.org/10.1215/9780822394716.

Bestor, T. C. 1989. *Neighborhood Tokyo*. Redwood City, CA: Stanford University Press.

Biltgen, Patrick and Stephen Ryan. 2016. *Activity-Based Intelligence: Principles and applications* (1st ed.). Norwood, MA: Artech House.

Bluteau, J. M. 2021. 'Gazing on invisible men: Introducing the gallery gaze to establish that (in)visibility is in the eye of the beholder at Westminster Menswear Archive', *Journal of Material Culture*. https://doi.org/10.1177/13591835211066821.

Borovoy, A. 2005. *The Too-Good Wife*. Berkeley, CA: University of California Press.

Borovoy, A. and K. Ghodsee. 2012. 'Decentering agency in feminist theory: Recuperating the family as a social project', *Women's Studies International Forum* 35(3): 153–165. https://doi.org/10.1016/j.wsif.2012.03.003.

Bourdieu, P. 1977. *Outline of a Theory of Practice*. Translated by Richard Nice. Cambridge: Cambridge University Press.

Brandtzæg, P. B. 2012. 'Social networking sites: Their users and social implications. A longitudinal study', *Journal of Computer-Mediated Communication* 17 (4): 467–88.

Brighenti, A. 2007. 'Visibility: A category for the social sciences', *Current Sociology* 55(3): 323–42. https://doi.org/10.1177/0011392107076079.

Brinton, M. C. and E. Oh. 2019. 'Babies, work, or both? Highly educated women's employment and fertility in East Asia', *American Journal of Sociology* 125(1): 105–40. https://doi.org/10.1086/704369.

Brown, K. 1966. '*Dōzoku* and the ideology of descent in rural Japan', *American Anthropologist* 68(5): 1129–51. https://doi.org/10.1525/aa.1966.68.5.02a00020.

Brumann, C. 2012. *Tradition, Democracy and the Townscape of Kyoto: Claiming a right to the past*. Abingdon, Oxon; New York, NY: Routledge. https://doi.org/10.4324/9780203135846.

Bucher, T. 2012. 'Want to be on the top? Algorithmic power and the threat of invisibility on Facebook', *New Media & Society* 14(7): 1164–80. https://doi.org/10.1177/1461444812440159.

Byford, S. 2012. 'Docomo's raku raku smartphone F12-D reinvents Android 4.0 for seniors (hands-on)', *The Verge*, 16 May 2012. https://www.theverge.com/2012/5/16/3023710/docomo-raku-raku-smartphone-f12-d-android-4-0-seniors.

Cabinet Office of the Japanese Government. 2014. 'Awareness survey on the everyday lives of the elderly in 2014'. http://www8.cao.go.jp/kourei/ishiki/h26/sougou/zentai/index.html (in Japanese).

Cabinet Public Relations Office. 2017. *Jinsei 100 nen jidai kousou*. https://www.kantei.go.jp/jp/headline/ichiokusoukatsuyaku/jinsei100.html (in Japanese).

Cacioppo, J. T. and L. C. Hawkley. 2009. 'Perceived social isolation and cognition', *Trends in Cognitive Sciences* 13(10): 447–54.

Campbell, J. C. and N. Ikegami. 2000. 'Long-term care insurance comes to Japan: A major departure for Japan, this new program aims to be a comprehensive solution to the problem of caring for frail older people', *Health Affairs* 19(3): 26–39.

Causey, A. 2017. *Drawn to See: Drawing as an ethnographic method*. Toronto: University of Toronto Press.

Champion, T. and J. Shepherd. 2006. 'Demographic change in rural England'. In *The Ageing Countryside: The growing older population of rural England*, edited by L. Speakman and P. Lowe, 29–50. London: Age Concern.

Chan, H. H. and L. L. Thang. 2022. 'Active aging through later life and afterlife planning: *Shūkatsu* in a super-aged Japan', *Social Sciences* 11(1): 3.

Chapman, S. A. 2005. 'Theorizing about aging well: Constructing a narrative', *Canadian Journal on Aging/La Revue canadienne du vieillissement* 24: 9–18. https://doi.org/10.1353/cja.2005.0004.

Chau, C. 2021. 'Japan approves law raising retirement age to 70', *HRM Asia*, 6 April 2021. http://hrmasia.com/japan-approves-law-raising-retirement-age-to-70/.

Chia, A. 2019. 'The moral calculus of vocational passion in digital gaming', *Television & New Media* 20(8): 767–77.

Chopik, W. J. 2016. 'The benefits of social technology use among older adults are mediated by reduced loneliness', *Cyberpsychology, Behavior, and Social Networking* 19(9), 551–6.

Clammer, J. 2010. *Difference and modernity: Social theory and contemporary Japanese society*. Abingdon, Oxon; New York, NY: Routledge.

Coates, J. 2019. 'The cruel optimism of mobility', *Positions: Asia Critique* 27(3): 469–97. https://doi.org/10.1215/10679847-7539277.

Cohn, N. 2014. 'The architecture of visual narrative comprehension: The interaction of narrative structure and page layout in understanding comics', *Frontiers in Psychology* 5: 1–9. https://doi.org/10.3389/fpsyg.2014.00680.

Collier, J. 1957. 'Photography in anthropology: A report on two experiments', *American anthropologist* 59(5): 843–59.

Collins, S. G., M. Durington, P. Favero, K. Harper, A. Kenner and C. O'Donnell. 2017. 'Ethnographic apps/apps as ethnography', *Anthropology Now* 9(1): 102–18. https://doi.org/10.1080/19428200.2017.1291054.

Colloredo-Mansfeld, R. 2011. 'Space, line and story in the invention of an Andean aesthetic', *Journal of Material Culture* 16(1): 3–23. https://doi.org/10.1177/1359183510394945.

Cons, J., T. Middleton, P. Atkinson, S. Delamont, A. Cernat, J. W. Sakshaug and R. A. Williams. 2020. 'Research assistants and ethnographic fieldwork'. https://methods.sagepub.com/foundations/research-assistants-and-ethnographic-fieldwork.

Cook, E. E. 2013. 'Expectations of failure: Maturity and masculinity for freeters in contemporary Japan', *Social Science Japan Journal* 16(1): 29–43. https://doi.org/10.1093/ssjj/jys022.

Cook, E. E. 2016. *Reconstructing Adult Masculinities: Part-time work in contemporary Japan*. Abingdon, Oxon; New York, NY: Routledge. https://doi.org/10.4324/9781315692548.

Cook, E. E. 2018. 'Socialization to acting, feeling, and thinking as *Shakaijin*: New employee orientations in a Japanese company'. In *Japanese at Work: Politeness, power, and personae in Japanese workplace discourse*, edited by H. M. Cook and J. S. Shibamoto-Smith, 37–64. London: Springer International Publishing. https://doi.org/10.1007/978-3-319-63549-1_3.

Cooper, A. 2018. *The Other Greek: An introduction to Chinese and Japanese characters, their history and influence*. Leiden: Brill.

Cox, R. A., A. Irving and C. Wright. 2016. *Beyond Text?: Critical practices and sensory anthropology*. Manchester: Manchester University Press.

Crawford, M. 2021. 'Abe's womenomics policy, 2013–2020: Tokenism, gradualism, or failed strategy?', *Asia-Pacific Journal: Japan focus*: 1–16.

Cruz, E. G. and E. Ardèvol. 2013. 'Ethnography and the field in media(ted) studies: A practice theory approach', *Westminster Papers in Communication and Culture* 9(3): 27. https://doi.org/10.16997/wpcc.172.

Cruz, E. G. and A. Lehmuskallio. 2016. *Digital Photography and Everyday Life*. Abingdon, Oxon; New York, NY: Routledge.

Czaja, S., M. Gold, L. J. Bain, J. A. Hendrix and M. C. Carrillo. 2017. 'Potential roles of digital technologies in clinical trials', *Alzheimers & Dementia* 13(9): 1075–6.

Danely, J. 2013. 'Temporality, spirituality and the life course in an aging Japan'. In *Transitions and Transformations: Cultural perspectives on aging and the life course* 1, edited by C. Lynch and J. Danely, 107. Oxford: Berghahn Books.

Danely, J. 2015a. *Aging and Loss: Mourning and maturity in contemporary Japan*. New Brunswick, NJ: Rutgers University Press. https://doi.org/10.36019/9780813565187.

Danely, J. 2015b. 'Of technoscapes and elderscapes: Editor's commentary on the special issue "Aging the Technoscape"', *Anthropology & Aging* 36(2): 110–11. https://doi.org/10.5195/aa.2015.121.

Danely, J. 2019. '"I don't want to live too long!": Successful aging and the failure of longevity in Japan'. In *Interrogating the Neoliberal Lifecycle: The limits of success*, edited by B. Clack and M. Paule, 189–212. New York, NY: Springer International Publishing. https://doi.org/10.1007/978-3-030-00770-6_9.

Danely, J. 2022. 'What older prisoners teach us about care and justice in an aging world', *Anthropology & Aging* 43(1): 58–65. https://doi.org/10.5195/aa.2022.395.

Danely, J. 2023. 'In the shadows of gratitude: On mooded spaces of vulnerability and care', *Ethos* 2023. 1–17. https://doi.org/10.1111/etho.12414.

Danely, J. and C. Lynch. 2013. 'Introduction: Transitions and transformations: Paradigms, perspectives, and possibilities'. In *Transitions and Transformations: Cultural perspectives on aging and the life course*, edited by Jason Danely and Caitrin Lynch, 3–20. Oxford: Berghahn Books.

Danesi, M. 2017. *The Semiotics of Emoji: The rise of visual language in the age of the internet*. London: Bloomsbury.

Daniels, I. 2001. 'The fame of Miyajima: Spirituality, commodification and the tourist trade of souvenirs in Japan'. Unpublished doctoral thesis, University of London.

Daniels, I. 2008. 'Japanese homes inside out', *Home Cultures* 5(2): 115–39. https://doi.org/10.2752/174063108X333155.

Daniels, I. 2010. *The Japanese House: Material culture in the modern home*. Abingdon, Oxon; New York, NY: Routledge.

Daniels, I. 2015. 'Feeling at home in contemporary Japan: Space, atmosphere and intimacy', *Emotion, Space and Society* 15: 47–55.

Daniels, I. 2022. Personal communication 14.02.2022.

Daniels, I. 2024. '"Even a piece of paper has two sides": multi-scalar cosmologies of Japanese New Year cards', *Journal of the Royal Anthropological Institute*. https://doi.org/10.1111/1467-9655.14099.

Dattatreyan, E. and I. Marrero-Guillamón. 2019. 'Introduction: Multimodal anthropology and the politics of invention', *American Anthropologist* 121(1): 220–8. https://doi.org/10.1111/aman.13183.

Davies, R. 2019. *Extreme Economies: Survival, failure, future lessons from the world's limits*. New York, NY: Random House.

De Mente, B. L. 2005. *Japan Unmasked: The character and culture of the Japanese*. Singapore: Tuttle Publishing.

De Togni, G. 2021. *Fall-out from Fukushima: Nuclear evacuees seeking compensation and legal protection after the triple meltdown*. Abingdon, Oxon; New York, NY: Routledge.

Depper, A. and P. D. Howe. 2017. '"Are we fit yet?" English adolescent girls' experiences of health and fitness apps', *Health Sociology Review* 26: 98–112.

Dix, B. and R. Kaur. 2019. 'Drawing-writing culture: The truth-fiction spectrum of an ethno-graphic novel on the Sri Lankan civil war and migration', *Visual Anthropology Review* 35(1): 76–111. https://doi.org/10.1111/var.12172.

Dobashi, S. 2005. 'Gendered usage of *keitai* in domestic contexts'. In *Personal, Portable, Pedestrian: Mobile phones in Japanese life*, edited by M. Ito, D. Okabe and M. Matsuda. Cambridge, MA: MIT Press.

Doi, T. 1971. *Amae no kòzò – The anatomy of dependence*. Tokyo: Kodansha.

Dore, R. P. 1973a. *British Factory–Japanese Factory: The origins of national diversity in industrial relations*. London: George Allen & Unwin.

Dore, R. P. 1973b. *City Life in Japan: A study of a Tokyo ward*. Berkeley, CA: University of California Press.

Dore, R. P. 1978. *Shinohata: A portrait of a Japanese village*. London: Allen Lane.

Dresner, E. and S. C. Herring. 2010. 'Functions of the nonverbal in CMC: Emoticons and illocutionary force', *Communication Theory* 20(3): 249–68. https://doi.org/10.1111/j.1468-2885.2010.01362.x.

Driessen, A. 2018. 'Pleasure and dementia: On becoming an appreciating subject', *The Cambridge Journal of Anthropology* 36(1): 23–39. https://doi.org/10.3167/cja.2018.360103.

Duque, M. 2022. *Ageing with Smartphones in Brazil: A work in progress*. London: UCL Press.

Edmunds, J. and B. S. Turner (eds). 2002. *Generational Consciousness, Narrative, and Politics*. Washington, DC: Rowman & Littlefield Publishers.

Edwards, E. 1992. *Anthropology and Photography, 1860–1920*. London: Royal Anthropological Institute of Great Britain and Ireland.

Eiraku, M. 2019. 'Japan's "hikikomori" are growing older', NHK WORLD-JAPAN News. NHK WORLD. https://www3.nhk.or.jp/nhkworld/en/news/backstories/464/.

Elder Jr, G. H. 1998. 'The life course as developmental theory', *Child Development* 69(1): 1–12.

Embree, J. F. 1946. *A Japanese Village, Suye mura*. London: K. Paul, Trench, Trubner.

Emde, R. N. 1992. '*Amae*, intimacy, and the early moral self', *Infant Mental Health Journal* 13(1):34–42.https://doi.org/10.1002/1097-0355(199221)13:1<34::AID-IMHJ2280130107>3.0.CO;2-S.

Fang, Y., A. K. C. Chau, A. Wong, H. H. Fung and J. J. Woo. 2018. 'Information and communicative technology use enhances psychological well-being of older adults: The roles of age, social connectedness, and frailty status', *Aging & Mental Health* 22(11): 1516–24. https://doi.org/10.1080/13607863.2017.1358354.

Favero, P. S. H. 2018. *The Present Image: Visible stories in a digital habitat*. London: Springer International Publishing. https://doi.org/10.1007/978-3-319-69499-3_1.

Favero, P. S. H. and E. Theunissen. 2018. 'With the smartphone as field assistant: Designing, making, and testing ethnoAlly, a multimodal tool for conducting serendipitous ethnography in a multisensory world', *American Anthropologist* 120(1): 163–7. https://doi.org/10.1111/aman.12999.

Fealy, G., M. McNamara, P. Treacy and I. Lyons. 2012. 'Constructing ageing and age identities: A case study of newspaper discourses', *Ageing & Society* 32(1): 85–102. https://doi.org/10.1017/S0144686X11000092.

Feldhoff, T. 2020. 'Structures and dynamics of Japan's urban-rural relationship'. In *Routledge Handbook of Contemporary Japan*. Abingdon, Oxon; New York, NY: Routledge.

Fire and Disaster Management Agency. 2021. 'Press release no. 161 of the 2011 Tohoku earthquake'. 総務省消防庁災害対策本部. https://www.fdma.go.jp/disaster/higashinihon/items/161.pdf.

Firth, R. 2020. 'Mutual aid, anarchist preparedness and COVID-19'. In *Coronavirus, Class and Mutual Aid in the United Kingdom*, edited by J. Preson and R. Firth, 57–111. London: Palgrave Macmillan. https://doi.org/10.1007/978-3-030-57714-8_4.

Foreman, K. M. 2017 [2008]. *The Gei of Geisha: Music, identity and meaning*. Abingdon, Oxon; New York, NY: Routledge. https://doi.org/10.4324/9781315086439.

Fotopoulou, A. and K. O'Riordan. 2016. 'Training to self-care: Fitness tracking, biopedagogy and the healthy consumer', *Health Sociology Review* 26: 54–68.

Foucault, M. 1977. *Discipline and Punish: The birth of the prison*. Translated by Alan Sheridan. London: Allen Lane.

Foucault, M. 1991. 'Governmentality'. In *The Foucault Effect: Studies in governmentality*, edited by G. Burchell, C. Gordon and P. Miller, 87–104. Chicago, IL: University of Chicago Press.

Foucault, M. 1998. *Foucault*. In *Aesthetics, Method, and Epistemology: Essential works of Foucault 1958–1984* (Vol. 2), edited by J. D. Faubion. New York: The New Press.

Francis, J. 2018. *Thanks to Facebook, getting old isn't that bad and I am not all alone in this world: An investigation of the effect of Facebook use on mattering and loneliness among elder orphans*. Doctoral thesis. East Lansing, MI: Michigan State University. https://doi.org/doi:10.25335/q7kj-ej23.

Fraser, N. 2011. 'Reflections: An interview with Nancy Fraser, interviewed by Amrita Chhachhi', *Development and Change* 42(1): 297–314.

Fraser, N. 2016. 'Capitalism's crisis of care', *Dissent* 63(4): 30–37. https://doi.org/10.1353/dss.2016.0071.

Fujita, K., Y. Takeuchi and R. Shaw. 2011. 'Voluntary self-help organization and fire volunteer for mountain disaster risk reduction in Reihoku area, Japan', *Asian Journal of Environment and Disaster Management* 3(3). http://rpsonline.com.sg/journals/101-ajedm/2011/0303/S1793924011000940.xml.

Fujiwara, K. 2012. 'Rethinking successful aging from the perspective of an aging Japanese statue of Jizō with replaceable heads', *Anthropology & Aging* 33(3): 104–11.

Fukunaga, R., Y. Abe, Y. Nakagawa, A. Koyama, N. Fujise and M. Ikeda. 2012. 'Living alone is associated with depression among the elderly in a rural community in Japan', *Psychogeriatrics* 12(3): 179–85. https://doi.org/10.1111/j.1479-8301.2012.00402.x.

Furst, M. 2017. 'Care and nutrition: Ethical issues: Exploring the moral nexus between caring and eating through natural history, anthropology and the ethics of care', *Relations. Beyond Anthropocentrism* 5: 103.

Gagné, N. O. 2018. '"Correcting capitalism": Changing metrics and meanings of work among Japanese employees', *Journal of Contemporary Asia* 48(1): 67–87. https://doi.org/10.1080/00472336.2017.1381984.

Gagné, N. O. 2020. 'Neoliberalism at work: Corporate reforms, subjectivity, and post-Toyotist affect in Japan', *Anthropological Theory* 20(4), 455–83. https://doi.org/10.1177/1463499618807294.

Gagné, N. O. 2021. *Reworking Japan: Changing men at work and play under neoliberalism*. Ithaca, NY: Cornell University Press.

Gandy Jr, O. H. 1989. 'The surveillance society: Information technology and bureaucratic social control', *Journal of Communication* 39(3): 61–76.

Garon, S. 2010. 'State and family in modern Japan: A historical perspective', *Economy and Society* 39(3): 317–336. https://doi.org/10.1080/03085147.2010.486214.

Garvey, P. and D. Miller. 2021. *Ageing with Smartphones in Ireland: When life becomes craft*. London: UCL Press. https://doi.org/10.14324/111.9781787359666.

Gatt, C. and T. Ingold. 2013. 'From description to correspondence: Anthropology in real time'. In *Design Anthropology: Theory and practice*, edited by W. Gunn, T. Otto and R.C. Smith, 139–58. London: Bloomsbury.

Gauntlett, D. 2007. *Creative Explorations: New approaches to identities and audiences*. Abingdon, Oxon; New York, NY: Routledge.

Germer, A., V. Mackie and U. Wöhr. 2014. *Gender, Nation and State in Modern Japan*. Abingdon, Oxon; New York, NY: Routledge.

Giele, J. Z. and G. H. Elder. 1998. *Methods of Life Course Research: Qualitative and quantitative approaches*. Thousand Oaks, CA: Sage.

Gill, R. 2014. 'Unspeakable Inequalities: Post feminism, entrepreneurial subjectivity, and the repudiation of sexism among cultural workers', *Social Politics: International Studies in Gender, State & Society* 21(4): 509–28.

Gilleard, C. and P. Higgs. 2002. 'The third age: Class, cohort or generation?', *Ageing & Society* 22(3): 369–82.

Gilleard, C. and P. Higgs. 2010. 'Aging without agency: Theorizing the fourth age', *Aging & Mental Health* 14(2): 121–8.

Gilleard, C. and P. Higgs. 2011. 'Frailty, disability and old age: A re-appraisal', *Health* 5 (5): 475–90.

Ginsburg, F. 1995. 'The parallax effect: The impact of aboriginal media on ethnographic film', *Visual Anthropology Review* 11(2): 64–76. https://doi.org/10.1525/var.1995.11.2.64.

Goffman, E. 1959. *The Presentation of Self in Everyday Life*. (Anchor Books edition). Garden City, NY: Doubleday.

Gómez Cruz, Edgar and Asko Lehmuskallio. 2016. *Digital Photography and Everyday Life*. Abingdon, Oxon; New York, NY: Routledge.

Goodman, R. 2005. 'Making majority culture'. In *A Companion to the Anthropology of Japan*, edited by J. E. Robertson, 59–72. Hoboken, NJ: Wiley-Blackwell.

Goodman, R. and K. Refsing (eds). 1992. *Ideology and Practice in Modern Japan*. Abingdon, Oxon; New York, NY: Routledge.

Goodwin, M. H. 2015. 'A care-full look at language, gender, and embodied intimacy'. In *Shifting Visions: Gender and Discourses*, edited by A. Jule, 27–48. Newcastle upon Tyne: Cambridge Scholars Publishing.

Gordon, M. 1996. *What Makes Life Worth Living? How Japanese and Americans make sense of their worlds*. Oakland, CA: University of California Press.

Grady, A., S. Yoong, R. Sutherland, H. Lee, N. Nathan and L. Wolfenden. 2018. 'Improving the public health impact of eHealth and mHealth interventions', *Australian and New Zealand Journal of Public Health* 42(2).

Grøn, L. and C. Mattingly. 2022. 'Introduction: Imagistic inquiries: Old age, intimate others, and care'. In *Imagistic Care: Growing old in a precarious world*, edited by Cheryl Mattingly and Lone Grøn, 1–30. New York, NY: Fordham University Press; online edn, Fordham Scholarship Online, 19 January 2023. https://doi.org/10.5422/fordham/9780823299645.003.0001.

Gubrium, A. and K. Harper. 2016. *Participatory Visual and Digital Methods*. Abingdon, Oxon; New York, NY: Routledge.

Gupta, A. 2014. 'Authorship, research assistants and the ethnographic field', *Ethnography* 15(3): 394–400.

Gupta, A. and J. Ferguson. 1997. *Culture, Power, Place: Explorations in critical anthropology*. Durham, NC: Duke University Press.

Haapio-Kirk, L. 2020. 'Staying connected: Coronavirus in Japan', *Entanglements* 3(2): 69–78. https://entanglementsjournal.wordpress.com/staying-connected-coronavirus-in-japan/.

Haapio-Kirk, L. Forthcoming. 'Digital repertoires of care in Japan: a participatory visual approach'. In *Embedded and Everyday Technology: Digital repertoires in an ageing society*, edited by R. Hänninen, S. Taipale and L. Haapio-Kirk. London: UCL Press.

Haapio-Kirk, L. and M. Ito. 2022. 'Turning points'. In *Ethno-graphic Collaborations: Crossing borders with multimodal illustration*, edited by L. Haapio-Kirk. *Trajectoria*, Volume 4. Osaka: National Museum of Ethnology. https://trajectoria.minpaku.ac.jp/articles/2022/vol03/01_1.html.

Haapio-Kirk, L. and L. Sasaki. Forthcoming. 'Rural Japan in transition: Mutual assistance, ageing, and the smartphone revolution'. In *Handbook of Rural Japan*, edited by S. Klien and P. Hansen. Amsterdam: Amsterdam University Press.

Haapio-Kirk, L., L. Sasaki and Y. Kimura. 2024. 'From "datafication" to socialisation: Rethinking self-tracking in Japan'. In *An anthropological approach to mHealth*, edited by C. Hawkins, D. Miller and P. Awondo. London: UCL Press.

Hall, E. T. 1989. *Beyond Culture*. New York: Anchor Books.

Hallam, E. 2009. 'Anatomists' ways of seeing and knowing'. In *Fieldnotes and Sketchbooks: Challenging the boundaries between descriptions and processes of describing*, edited by W. Gunn, 69–108. Hamburg: Peter Lang.

Hamaguchi, E. 1982. *Kanjin shugi no shakai Nihon (Japan as a contextual society)*. Tokyo: Toyo Keizai Shinposha.

Han, C. 2018. 'Precarity, precariousness, and vulnerability', *Annual Review of Anthropology* 47, 331–43.

Hänninen, R., L. Pajula, V. Korpela and S. Taipale. 2021. 'Individual and shared digital repertoires – older adults managing digital services', *Information, Communication & Society* 26(3): 568–83. https://doi.org/10.1080/1369118X.2021.1954976.

Hansen, P. and S. Klien. 2022. Special issue. 'Exploring rural Japan as heterotopia', *Asian Anthropology* 21(1): 1–9. https://doi.org/10.1080/1683478X.2021.2015110.

Hardt, M. and A. Negri. 2001. *Empire*. Cambridge, MA: Harvard University Press.

Hareven, T. 2003. *The Silk Weavers of Kyoto: Family and work in a changing traditional industry* Berkeley, CA: University of California Press. https://doi.org/10.1525/9780520935761.

Hasebrink, U. and H. Domeyer. 2012. 'Media repertoires as patterns of behaviour and as meaningful practices: A multimethod approach to media use in converging media environments', *Participations* 9(2): 757–79.

Hasebrink, U. and A. Hepp. 2017. 'How to research cross-media practices? Investigating media repertoires and media ensembles', *Convergence* 23(4): 362–77. https://doi.org/10.1177/1354856517700384.

Hata, H. and W. A. Smith. 1986. 'The vertical structure of Japanese society as a utopia', *Review of Japanese Culture and Society* 1(1): 92–109.

Headquarters for Japan's Economic Revitalization. 2019. *Action Plan of the Growth Strategy*. https://www.kantei.go.jp/jp/singi/keizaisaisei/pdf/ap2019en.pdf.

Heffernan, E., F. Murphy and J. Skinner. 2020. *Collaborations: Anthropology in a neoliberal age*. Abingdon, Oxon; New York, NY: Routledge.

Hendry, J. 1981. *Marriage in Changing Japan: Community and society*. London: Croom Helm. https://doi.org/10.4324/9780203841334.

Hendry, J. 1989. 'To wrap or not to wrap: Politeness and penetration in ethnographic inquiry', *Man* 24(4): 620–35. https://doi.org/10.2307/2804291.

Hendry, J. 1993. *Wrapping Culture: Politeness, presentation and power in Japan and other societies*. Oxford: Oxford University Press.

Hendry, J. 2017. *An Anthropological Lifetime in Japan*. Leiden: Brill.

Higgs, P. and C. Gilleard. 2015. *Rethinking Old Age: Theorising the fourth age*. London: Bloomsbury.

Hine, C. 2020. *Ethnography for the Internet: Embedded, embodied and everyday*. Abingdon, Oxon; New York, NY: Routledge.

Hirai, M. 2011. 'Distribution and migration of elderly population'. In *Population Geography of Contemporary Japan*, edited by Y. Ishikawa, T. Inoue and Y. Tahara. Tokyo: Kokon Shoin.

Hirayama, Y. and M. Izuhara. 2018. *Housing in Post-Growth Society: Japan on the edge of social transition*. Abingdon, Oxon; New York, NY: Routledge.

Hirayama, Y. and R. Ronald. 2008. 'Baby-boomers, baby-busters and the lost generation: Generational fractures in Japan's homeowner society', *Urban Policy and Research* 26(3): 325–42. https://doi.org/10.1080/08111140802301773.

Hjorth, L. and K. Y. Kim. 2011. 'The mourning after: A case study of social media in the 3.11 earthquake disaster in Japan', *Television & New Media* 12(6): 552–9. https://doi.org/10.1177/1527476411418351.

Hjorth, L., K. Ohashi, J. Sinanan, H. Horst, S. Pink, F. Kato and B. Zhou. 2020. *Digital Media Practices in Households: Kinship through data*. Amsterdam: Amsterdam University Press. https://library.oapen.org/handle/20.500.12657/40043.

Hochschild, A. R. 1979. 'Emotion work, feeling rules, and social structure', *American Journal of Sociology* 85(3): 551–75. https://doi.org/10.1086/227049.

Hockey, J. and M. Forsey. 2012. 'Ethnography is not participant observation: Reflections on the interview as participatory qualitative research'. In *The Interview: An Ethnographic Approach*, edited by J. Skinner, 69–88. Oxford; New York, NY: Berg.

Hoey, B. A. 2010. 'Place for personhood: Individual and local character in lifestyle migration', *City & Society* 22(2): 237–61. https://doi.org/10.1111/j.1548-744X.2010.01041.x.

Hogan, S. 2017. 'Working across disciplines: Using visual methods in participatory frameworks.' In *Theoretical Scholarship and Applied Practice*, edited by S. Pink, V. Fors and T. O'Dell, 142–66. London: Berghahn Books.

Hong, R. 2022. *Passionate Work: Endurance after the good life*. Durham, NC: Duke University Press.

Horiguchi, S. 2011. 'Hikikomori: How private isolation caught the public eye'. In *A Sociology of Japanese Youth*, edited by R. Goodman, I. Yuki and T. Toivonen, 142–58. Abingdon, Oxon; New York, NY: Routledge.

Horst, H. and D. Miller. 2006. *The Cell Phone: An anthropology of communication*. Abingdon, Oxon; New York, NY: Berg.

Horst, H. A. and D. Miller. 2020. *Digital Anthropology*. Abingdon, Oxon; New York, NY: Routledge.

Human Mortality Database 2021. University of California, Berkeley (USA), and Max Planck Institute for Demographic Research (Germany). www.mortality.org or www.humanmortality.de (data downloaded on 7.3.2021).

Hutchison, E. D. 2018. *Dimensions of Human Behavior: The changing life course*. Thousand Oaks, CA: Sage.

Hutto, C. and C. Bell. 2014. 'Social media gerontology: Understanding social media usage among a unique and expanding community of users', *System Sciences (HICSS), 47th Hawai'i International Conference, 2014*. Piscataway, NJ: IEEE.

Ikeda, N., E. Saito, N. Kondo, M. Inoue, S. Ikeda, T. Satoh, K. Wada, A. Stickley, K. Katanoda, T. Mizoue and M. Noda. 2011. 'What has made the population of Japan healthy?', *The Lancet* 378(9796): 1094–105.

Ingold, T. 2000. *The Perception of the Environment: Essays on livelihood, dwelling and skill*. Abingdon, Oxon; New York, NY: Routledge.

Ingold, T. 2007. *Lines: A brief history*. Abingdon, Oxon; New York, NY: Routledge.

Ingold, T. 2019. 'Art and anthropology for a sustainable world', *Journal of the Royal Anthropological Institute* 25(4): 659–75.

Ishikawa, M. 2020. 'Media portrayals of ageing baby boomers in Japan and Finland'. Doctoral thesis. *Publications of the Faculty of Social Sciences* 159 (2020), Social and Public Policy. Helsinki: University of Helsinki.

Ito, M. 2005. 'Mobile phones, Japanese youth, and the re-placement of social contact'. In *Mobile Communications: Re-negotiation of the social sphere*, edited by S. Woolgar, R. Ling and P. E. Pedersen, 131–48. London: Springer.

Ito, M. and D. Okabe. 2005. 'Technosocial situations: Emergent structurings of mobile email use'. In *Personal, Portable, Pedestrian: Mobile phones in Japanese life*, edited by M. Ito, D. Okabe and M. Matsuda. Cambridge, MA: MIT Press.

Ito, M., D. Okabe and M. Matsuda (eds). 2005. *Personal, Portable, Pedestrian: Mobile phones in Japanese life*. Cambridge, MA: MIT Press.

Ivy, Marilyn. 1995. *Discourses of the Vanishing: Modernity, phantasm, Japan*. Chicago, IL; London: University of Chicago Press.

Izlar, J. 2019. 'Radical social welfare and anti-authoritarian mutual aid', *Critical and Radical Social Work* 7(3). https://doi.org/10.1332/204986019X15687131179624.

Izuhara, M. 2015. 'Life-course diversity, housing choices and constraints for women of the 'lost' generation in Japan', *Housing Studies* 30(1): 60–77. https://doi.org/10.1080/02673037.2014.933780.

Jackson, Michael. 2000. *At Home in the World*. Durham, NC: Duke University Press.

Jackson, M. 2005. *Existential Anthropology: Events, exigencies, and effects*. New York: Berghahn.

Jackson, M. 2012. *Between One and One Another*. Berkeley, CA: University of California Press. https://doi.org/10.1525/9780520951914.

Jackson, M. and A. Piette. (eds). 2015. *What is Existential Anthropology?* New York: Berghahn Books.

Janoschka, M. and H. Haas. 2013. *Contested Spatialities, Lifestyle Migration and Residential Tourism*. Abingdon, Oxon; New York, NY: Routledge.

Japan Pension Service. 2024. *National Pension System*. Retrieved 8 March 2024, from https://www.nenkin.go.jp/international/japanese-system/nationalpension/nationalpension.html.

Japanese Ministry of Public Management, Home Affairs, Posts and Telecommunications. 2002. *White Paper: Information and Communications in Japan*. Tokyo: National Printing Bureau.

Johnson, C. 1982. *MITI and the Japanese Miracle: The growth of industrial policy, 1925–1975*. Redwood City, CA: Stanford University Press.

Jones, R. and H. Seitani. 2019. 'Meeting fiscal challenges in Japan's rapidly ageing society', *OECD Economics Department Working Papers*, No. 1569. Paris: OECD Publishing. https://doi.org/10.1787/7a7f4973-en.

Jurgenson, N. 2019. *The Social Photo: On photography and social media*. London: Verso.

Kanayama, T. 2003. 'Ethnographic research on the experience of Japanese elderly people online', *New Media & Society* 5(2): 267–88. https://doi.org/10.1177/1461444803005002007.

Kano, A. 2016. *Japanese Feminist Debates: A century of contention on sex, love, and labor*. Honolulu: University of Hawai'i Press. https://doi.org/10.1515/9780824855833.

Katanuma, M. 2020. 'Japan's population crisis is pushing more women into poverty', 11 January 2020. https://www.bloomberg.com/news/articles/2020-01-11/japan-s-population-crisis-ispushing-more-women-into-poverty.

Kavedžija, I. 2015a. 'The good life in balance: Insights from aging Japan', *HAU: Journal of Ethnographic Theory* 5(3): 135–56. https://doi.org/10.14318/hau5.3.008.

Kavedžija, I. 2015b. 'Frail, independent, involved? Care and the category of the elderly in Japan', *Anthropology & Aging* 36(1): 62–81.

Kavedžija, I. 2016. 'The age of decline? Anxieties about ageing in Japan', *Ethnos* 81(2): 214–37. https://doi.org/10.1080/00141844.2014.911769.

Kavedžija, I. 2018. 'Of manners and hedgehogs: Building closeness by maintaining distance', *The Australian Journal of Anthropology* 29(2): 146–57.

Kavedžija, I. 2019. *Making Meaningful Lives: Tales from an aging Japan*. Philadelphia, PA: University of Pennsylvania Press.

Kavedžija, I. 2021. *The Process of Wellbeing: Conviviality, care, creativity. Elements in psychology and culture*. Cambridge: Cambridge University Press. https://doi.org/10.1017/9781108935616.

Kazuo, Y. 2015. 'The effect of the baby boomer generation on Japan', *JAPAN SPOTLIGHT* 500: 10–11.

Kelly, W. W. 1990. 'Regional Japan: The price of prosperity and the benefits of dependency', *Daedalus* 119(3): 209–27.

Kelly, W. W. 1993a. 'Finding a place in metropolitan Japan: Ideologies, institutions, and everyday life'. In *Postwar Japan as History*, edited by Andrew Gordon, 189–238, Berkeley, CA: University of California Press.

Kelly, W. W. 1993b. Review of 'Changing Japanese suburbia: A study of two present-day localities' by E. Ben-Ari, *Monumenta Nipponica* 48(3): 401–3. https://doi.org/10.2307/2385140.

Kennedy, H., T. Poell and J. van Dijck. 2015. 'Introduction: "Data and agency"', *Big Data Society* 2. https://doi.org/10.1177/2053951715621569.

Kim, K. Y. 2017. 'Keitai in Japan'. In *Routledge Handbook of Japanese Media*, edited By F. Darling-Wolf, 308–20. Abingdon, Oxon; New York, NY: Routledge.

Kimura, Y., T. Wada, M. Ishine et al. 2008. 'Food diversity is closely associated with activities of daily living, depression, and quality of life in community-dwelling elderly people', *Journal of the American Geriatric Society* 57: 922–4.

Kimura, Y., T. Wada, K. Okumiya, Y. Ishimoto, E. Fukutomi, Y. Kasahara, W. Chen, R. Sakamoto, M. Fujisawa, K. Otsuka and K. Matsubayashi. 2012. 'Eating alone among community-dwelling Japanese elderly: Association with depression and food diversity', *The Journal of Nutrition, Health & Aging* 16(8): 728–31.

Kitayama, S., M. Karasawa, K. Curhan, C. Ryff and H. Markus. 2010. 'Independence and interdependence predict health and wellbeing: Divergent patterns in the United States and Japan', *Frontiers in Psychology*, 1. https://www.frontiersin.org/articles/10.3389/fpsyg.2010.00163.

Kleinman, A. 2009. 'Caregiving: The odyssey of becoming more human', *The Lancet 373*(9660): 292–3. https://doi.org/10.1016/S0140-6736(09)60087-8.

Klien, S. 2015. 'Young urban migrants in the Japanese countryside between self-realisation and slow life'. In *Sustainability in Contemporary Rural Japan: Challenges and opportunities*, edited by S. Assmann, 95–107. Abingdon, Oxon; New York, NY: Routledge.

Klien, S. 2016. 'Reinventing Ishinomaki, Reinventing Japan? Creative networks, alternative lifestyles and the search for quality of life in post-growth Japan', *Japanese Studies* 36(1): 39–60. https://doi.org/10.1080/10371397.2016.1148555.

Klien, S. 2019. 'Entrepreneurial selves, governmentality and lifestyle migrants in rural Japan', *Asian Anthropology* 18(2): 75–90. https://doi.org/10.1080/1683478X.2019.1572946.

Klien, S. 2020. '"Life could not be better since I left Japan!": Transnational mobility of Japanese individuals to Europe and the post-Fordist quest for subjective well-being outside Japan'. In *New Frontiers in Japanese Studies*, edited by A. Ogawa and P. Seaton, 194–206. Abingdon, Oxon; New York, NY: Routledge.

Knight. J. 1998. 'The second life of trees: Family forestry in upland Japan'. In *The Social Life of Trees: Anthropological perspectives on tree symbolism*, edited by Laura M. Rival, 197–220. Materializing Culture. Oxford: Berg.

Knight, J. 2003. 'Organic farming settlers in Kumano'. In *Farmers and Village Life in Twentieth-century Japan*, edited by A. Waswo and N. Yoshiaki, 267–284. Abingdon, Oxon: Taylor and Francis.

Knight, J. and J. W. Traphagan. 2003. 'The study of the family in Japan: Integrating anthropological and demographic approaches'. In *Demographic Change and the Family in Japan's Aging Society*, edited by J. W. Traphagan and J. Knight, 3–23. Albany, NY: Suny Press.

Kobayashi, E., I. Sugawara, T. Fukaya, S. Okamoto and J. Liang. 2022. 'Retirement and social activities in Japan: Does age moderate the association?', *Research on Aging* 44(2): 144–55. https://doi.org/10.1177/01640275211005185.

Kōchi Prefecture Tosa-Chō, home page. 2020. *Tosa Chō*. https://www.town.tosa.Kōchi.jp/. Accessed 26 August 2020.

Kojima, G., S. Iliffe, Y. Taniguchi, H. Shimada, H. Rakugi and K. Walters. 2017. 'Prevalence of frailty in Japan: A systematic review and meta-analysis', *Journal of Epidemiology* 27(8): 347–53. https://doi.org/10.1016/j.je.2016.09.008.

Komori, N. 2007. 'The "hidden" history of accounting in Japan: A historical examination of the relationship between Japanese women and accounting', *Accounting History* 12(3): 329–58. https://doi.org/10.1177/1032373207079037.

Kondo, D. K. 1990. *Crafting Selves: Power, gender, and discourses of identity in a Japanese workplace*. Chicago, IL: University of Chicago Press.

Korpela, M. 2014. 'Lifestyle of freedom? Individualism and lifestyle migration'. In *Understanding Lifestyle Migration: Theoretical approaches to migration and the quest for a better way of life*, edited by M. Benson and N. Osbaldiston, 27–46. London: Palgrave Macmillan UK. https://doi.org/10.1057/9781137328670_2.

Kress, G. 2003. *Literacy in the New Media Age*. Abingdon, Oxon; New York, NY: Routledge. https://doi.org/10.4324/9780203299234.

Kübler-Ross, E. and D. Kessler. 2005. *On Grief and Grieving: Finding the meaning of grief through the five stages of loss*. New York, NY: Simon and Schuster.

Kudo, S. and M. Yarime. 2013. 'Divergence of the sustaining and marginalising communities in the process of rural aging: A case study of Yurihonjo-shi, Akita, Japan', *Sustainability Science* 8(4): 491–513. https://doi.org/10.1007/s11625-012-0197-x.

Kumano, M. 2018. 'On the concept of well-being in Japan: Feeling *shiawase* as hedonic well-being and feeling *ikigai* as eudaimonic well-being', *Applied Research in Quality of Life* 13(2): 419–33. https://doi.org/10.1007/s11482-017-9532-9.

Kuwayama, T. 2001. 'The discourse of Ie (family) in Japan's cultural identity and nationalism: A critique', *Japanese Review of Cultural Anthropology* 2: 3–37.

Kyoto City official website 2016. *Ikigai* Measures. https://www.city.kyoto.lg.jp/hokenfukushi/page/0000207602.html. Accessed 30 December 2023.

Kyoto City official website 2021. https://www.city.kyoto.lg.jp/index.html. Accessed 25 May 2021.

Lachman, M. E. 2002. *Handbook of Midlife Development*. Hoboken, NJ: John Wiley.

Lachowicz, K. and J. Donaghey. 2022. 'Mutual aid versus volunteerism: Autonomous PPE production in the Covid-19 pandemic crisis', *Capital & Class* 46(3): 427–47. https://doi.org/10.1177/03098168211057686.

Lamb, S. 2014. 'Permanent personhood or meaningful decline? Toward a critical anthropology of successful aging', *Journal of Aging Studies* 29: 41–52. https://doi.org/10.1016/j.jaging.2013.12.006.

Lamb, Sarah (ed.). 2017. *Successful Aging as a Contemporary Obsession: Global perspectives*. New Brunswick, NJ: Rutgers University Press.

Lanzeni, D. and K. Waltorp. 2018. 'Interventions, design, and the many modalities of future-oriented anthropology'. London: 4th International Conference of the Royal Anthropological Institute.

Lapenta, F. 2011. 'Geomedia: On location-based media, the changing status of collective image production and the emergence of social navigation systems', *Visual Studies* 26(1): 14–24. https://doi.org/10.1080/1472586X.2011.548485.

Laslett, P. 1991. *A Fresh Map of Life: The emergence of the third age*. Cambridge, MA: Harvard University Press.

Lasén, A. and E. Gómez-Cruz. 2009. 'Digital photography and picture sharing: Redefining the public/private divide', *Knowledge, Technology & Policy* 22(3): 205–15. https://doi.org/10.1007/s12130-009-9086-8.

Leach, E. R. 1961. *Pul Eliya: A village in Ceylon*. Cambridge: Cambridge University Press.

Lebra, T. 1976. *Japanese Patterns of Behavior*. Honolulu: University of Hawai'i Press.

Lebra, T. S. 2004. *The Japanese Self in Cultural Logic*. Honolulu: University of Hawai'i Press. https://muse.jhu.edu/book/8167.

Lebra, T. S. and W. P. Lebra. 1986. *Japanese Culture and Behavior: Selected readings*. Honolulu: University of Hawai'i Press.

Leivestad, H. H. 2020. *Caravans: Lives on wheels in contemporary Europe*. Abingdon, Oxon; New York, NY: Routledge. https://doi.org/10.4324/9781003084877.

Lim, S. S. 2015. 'On stickers and communicative fluidity in social media', *Social Media + Society* 1(1). https://doi.org/10.1177/2056305115578137.

Lindén, L. and D. Lydahl. 2021. 'Editorial: Care in STS', *Nordic Journal of Science and Technology Studies*, 3–12. https://doi.org/10.5324/njsts.v9i1.4000.

Lindtner, S., J. Chen, G. R. Hayes and P. Dourish. 2011. 'Towards a framework of publics: Re-encountering media sharing and its user', *ACM Transactions on Computer-Human Interaction* 18(2): 5:1–5:23. https://doi.org/10.1145/1970378.1970379.

Linecorp. 2019. '[Global] LINE announces custom stickers – create your own stickers in minutes using popular LINE Characters', *LINE Corporation*. Retrieved from https://linecorp.com/en/pr/news/en/2019/2666.

Literat, I. 2013. '"A pencil for your thoughts": Participatory drawing as a visual research method with children and youth', *International Journal of Qualitative Methods* 12(1): 84–98. https://doi.org/10.1177/160940691301200143.

Littlejohn, A. 2021. 'Museums of themselves: Disaster, heritage, and disaster heritage in Tohoku', *Japan Forum* 33(4): 476–96. https://doi.org/10.1080/09555803.2020.1758751.

Lo, S.-K. 2008. 'The nonverbal communication functions of emoticons in computer-mediated communication', *CyberPsychology & Behavior* 11(5): 595–7. https://doi.org/10.1089/cpb.2007.0132.

Long, S. O., R. Campbell and C. Nishimura. 2009. 'Does it matter who cares? A comparison of daughters versus daughters-in-law in Japanese elder care', *Social Science Japan Journal* 12(1): 1–21. https://doi.org/10.1093/ssjj/jyn064.

Loos, E. and L. Ivan. 2018. 'Visual ageism in the media'. In *Contemporary Perspectives on Ageism*, edited by Liat Ayalon and Clemens Tesch-Römer (1st ed.), 163–76. Cham: Springer International Publishing. https://doi.org/10.1007/978-3-319-73820-8.

Love, B. 2013. 'Treasure hunts in rural Japan: Place making at the limits of sustainability', *American Anthropologist* 115(1): 112–24. https://doi.org/10.1111/j.1548-1433.2012.01539.x.

Luor, T., L. Wu, H.-P. Lu and Y.-H. Tao. 2010. 'The effect of emoticons in simplex and complex task-oriented communication: An empirical study of instant messaging', *Computers in Human Behavior* 26(5): 889–95. https://doi.org/10.1016/j.chb.2010.02.003.

Lupton, D. 2013a. 'The digitally engaged patient: Self-monitoring and self-care in the digital health era', *Social Theory Health* 11:256–70.

Lupton, D. 2013b. 'Quantifying the body: Monitoring and measuring health in the age of mHealth technologies', *Critical Public Health* 23(4): 393–403. https://doi.org/10.1080/09581596.2013.794931.

Lupton, D. 2014. 'Critical perspectives on digital health technologies', *Sociology Compass* 8: 1344–59. https://doi.org/10.1111/soc4.12226.

Lupton, D. 2017. *Digital Health: Critical and cross-disciplinary perspectives*. Abingdon, Oxon; New York, NY: Routledge.

Lupton, D. 2021. 'Self-tracking'. In *Information: Keywords*, edited by M. Kennerly, S. Frederick and J. Abel, 187–98. New York; Chichester, West Sussex: Columbia University Press. https://doi.org/10.7312/kenn19876-016.

Lutz, P. 2015. 'Multivalent moves in senior home care: From surveillance to care-valence', *Anthropology & Aging* 36(2): 145–63. https://doi.org/10.5195/aa.2015.105.

Lynch, C. 2020. 'Eldersourcing and reconceiving work as care'. In *The Cultural Context of Aging: Worldwide perspectives*, edited by J. Sokolovsky (4th ed.), 171. London: Bloomsbury.

MacDougall, D. 1978. 'Ethnographic film: Failure and promise', *Annual Review of Anthropology* 7(1): 405–25. https://doi.org/10.1146/annurev.an.07.100178.002201.

MacDougall, D. 2005. *The Corporeal Image: Film, ethnography, and the senses*. Princeton, NJ: Princeton University Press.

Macgillivray, I. K. 2005. 'Using cartoons to teach students about stereotypes and discrimination: One teacher's lessons from South Park', *Journal of Curriculum and Pedagogy* 2(1): 133–47.

Mackie, V. 1997. 'Introduction'. In *Creating Socialist Women in Japan: Gender, labour and activism, 1900–1937*, edited by V. Mackie, 1–21. Cambridge: Cambridge University Press. https://doi.org/10.1017/CBO9780511518270.001.

Mackie, V. 2003. *Feminism in Modern Japan: Citizenship, embodiment and sexuality*. Cambridge: Cambridge University Press.

Mackie, V. 2015. 'The crisis of care and the future of work in the Asia-Pacific Region', *Journal and Proceedings of the Royal Society of New South Wales*, vol. 148, nos. 457 & 458, 176–184.

Macnaughtan, H. 2015. 'Womenomics for Japan: Is the Abe policy for gendered employment viable in an era of precarity?' *The Asia-Pacific Journal: Japan Focus* 13(13): 4302.

Macnicol, J. 2015. *Neoliberalising Old Age*. Cambridge: Cambridge University Press.

Maldarelli, C. 2021. 'Why 10,000 steps a day isn't the secret to better health', *Popular Science*. https://www.popsci.com/story/health/10000-steps-evidence-study/.

Malinowski, B. 1923. 'The problem of meaning in primitive languages'. In *The Meaning of Meaning*, edited by C. K. Ogden and I. A. Richards, 146–52. London: Routledge & Kegan Paul.

Mannheim, K. 2023 [1928]. 'Das problem der generationen' (The Problem of Generations). In *Schriften zur Wirtschafts-und Kultursoziologie (Writings on Economic and Cultural Sociology)*, edited by Amalia Barboza und Klaus Lichtblau, 121–69. Wiesbaden: Springer Fachmedien Wiesbaden.

Manzenreiter, W., R. Lützeler and S. Polak-Rottmann (eds). 2020. *Japan's New Ruralities: Coping with decline in the periphery*. Abingdon, Oxon; New York, NY: Routledge.

Marcus, G. E. 1995. 'Ethnography in/of the world system: The emergence of multi-sited ethnography', *Annual Review of Anthropology* 24: 95–117.

Markham, A. N. 2013. 'Fieldwork in social media. What would Malinowski do?', *Qualitative Communication Research* 2(4): 434–46. https://doi.org/10.1525/qcr.2013.2.4.434.

Martela, F. and M. F. Steger. 2016. 'The three meanings of meaning in life: Distinguishing coherence, purpose, and significance', *The Journal of Positive Psychology* 11(5): 531–45.

Matanle, P. and A. Rausch. 2011. *Japan's Shrinking Regions in the 21st Century: Contemporary responses to depopulation and socioeconomic decline*. Cambridge: Cambridge University Press.

Mathews, G. 1996a. *What Makes Life Worth Living?: How Japanese and Americans make sense of their worlds*. Berkeley, CA: University of California Press.

Mathews, G. 1996b. 'The stuff of dreams, fading: *Ikigai* and "The Japanese Self"', *Ethos* 24(4): 718–47. https://www.jstor.org/stable/640520. Accessed 11 August 2021.

Mathews, G. and C. Izquierdo (eds). 2009. *Pursuits of Happiness: Well-Being in anthropological perspective*. Oxford; New York, NY: Berghahn.

Matsubayashi, K. and K. Okumiya. 2012. 'Field medicine: A new paradigm of geriatric medicine', *Geriatrics and Gerontology International* 12(1): 5–15.

Matsuda, M. 2009. 'Mum in the pocket'. In *Mobile Technologies: From telecommunication to media*, edited by G. Goggin and L. Hjorth, 62–72. Abingdon, Oxon; New York, NY: Routledge.

Maung, H. H. 2021. 'What's my age again? Age categories as interactive kinds', *History and Philosophy of the Life Sciences* 43(1): 36. https://doi.org/10.1007/s40656-021-00388-5.

Mayumi Ono. 2015. 'Commoditization of lifestyle migration: Japanese retirees in Malaysia', *Mobilities* 10(4): 609–27. https://doi.org/10.1080/17450101.2014.913868.

McCloud, S. 1993. *Understanding Comics: The invisible art*. Princeton, WI: Kitchen Sink Press.

McIntyre, N. 2009. 'Rethinking amenity migration: Integrating mobility, lifestyle and social-ecological systems', *Erde* 14(3): 229–50.

McKay, D. 2016. *An Archipelago of Care: Filipino migrants and global networks*. Bloomington, IN: Indiana University Press.

Merler, S. 2018. 'Abenomics, five years in: Has it worked?'. *World Economic Forum*. https://www.weforum.org/agenda/2018/01/abenomics-five-years-in-has-it-worked/.

Millar, K. M. 2017. 'Toward a critical politics of precarity', *Sociology Compass* 11: 6 e12483. https://doi.org/10.1111/soc4.12483.

Miller, D. 2015. 'Photography in the age of Snapchat', *Anthropology & Photography*, edited by the RAI Photography Committee. London: *Royal Anthropological Institute*. http://www.therai.org.uk/images/stories/photography/AnthandPhotoVol2.pdf.

Miller, D. 2016. *Social Media in an English Village*. London: UCL Press.

Miller, D. and J. Sinanan. 2014. *Webcam*. Cambridge; Oxford; Boston; New York: Polity Press.

Miller, D., E. Costa, N. Haynes, T. McDonald, R. Nicolescu, J. Sinanan, J. Spyer, S. Venkatraman and X. Wang. 2016. *How the World Changed Social Media*. London: UCL Press. https://doi.org/10.14324/111.9781910634493.

Miller, D. and L. Haapio-Kirk. 2020. 'Making things matter'. In *Lineages and Advancements in Material Culture Studies: Perspectives from UCL anthropology*, edited by T. Carroll, A. Walford and S. Walton, 146–57. Abingdon, Oxon; New York, NY: Routledge.

Miller, D. and H. Horst. 2020. 'The digital and the human: A prospectus for digital anthropology'. In *Digital Anthropology*, edited by H. Horst and D. Miller, 3–36. Abingdon, Oxon; New York, NY: Routledge.

Miller, D., L. Abed Rabho, P. Awondo, M. de Vries, M. Duque, P. Garvey, L. Haapio-Kirk, C. Hawkins, A. Otaegui, S. Walton and X. Wang. 2021. *The Global Smartphone: Beyond a youth technology*. London: UCL Press. https://doi.org/10.14324/111.9781787359611.

Miller, V. 2008. 'New media, networking and phatic culture', *Convergence* 14(4): 387–400. https://doi.org/10.1177/1354856508094659.

Mingas, M. 2020. 'Docomo's $40bn privatisation could bring lower phone tariffs for Japan', *Capacity Media*. https://www.capacitymedia.com/article/29otckqx2h8okpw2x57uo/news/docomos-40bn-privatisation-could-bring-lower-phone-tariffs-for-japan.

Minichiello, V., J. Browne and H. Kendig. 2000. 'Perceptions and consequences of ageism: Views of older people', *Ageing & Society* 20(3): 253–78. https://doi.org/10.1017/S0144686X99007710.

Ministry of Health, Labor and Welfare. 2020a. *Press Release* [in Japanese]. https://warp.da.ndl.go.jp/info:ndljp/pid/12764107/www.mhlw.go.jp/toukei/saikin/hw/life/life20/dl/life18-14.pdf.

Ministry of Health, Labor and Welfare. 2020b. *General Welfare and Labour. Overview of the system and the basic statistics*. https://www.mhlw.go.jp/english/wp/wp-hw13/dl/01e.pdf. Accessed 19 September 2021.

Ministry of Internal Affairs and Communication. 2020. *Population, Households, Sex, Age and Marital status*. https://www.stat.go.jp/english/data/kokusei/index.html.

Ministry of Internal Affairs and Communications. 2021. *System of Social and Demographic Statistics / Statistical Observations of Prefectures 2021 / Social Indicators by Prefecture*. https://www.e-stat.go.jp/en/stat-search/files?page=1&layout=datalist&toukei=00200502&tstat=00000 1149949&cycle=0&year=20210&month=0&tclass1=000001149950.

Miyamoto, K. 2014. 'Dankaisedairon no chushin mondai: Gendaishakairon no shiten kara' ['Central problems in studies on Japanese baby boomers (DANKAI Generation): From the viewpoint of theories of modern society'], *Momoyamagakuindaigaku shakaigaku ronshū* 48(1): 69–95.

Miyata, K., J. Boase and B. Wellman. 2008. 'The social effects of *keitai* and personal computer email in Japan'. In *Mobile Communications: Re-negotiation of the social sphere*, edited by R. Ling and P. E. Pedersen, 209–22. London: Springer. https://doi.org/10.7551/mitpress/9780262113120.003.0016.

Mizukoshi, K., F. Kohlbacher and C. Schimkowsky. 2016. 'Japan's *ikumen* discourse: Macro and micro perspectives on modern fatherhood', *Japan Forum* 28(2): 212–32. https://doi.org/10.1080/09555803.2015.1099558.

Mobile Society Research Institute. 2021. 'Mobile society white paper web version. Chapter 6: Living conditions of seniors and use of ICT'. https://www.moba-ken.jp/whitepaper/21_chap6.html. Accessed 3 February 2022.

Mock, J. 2012. 'The social impact of rural-urban shift: Some Akita examples'. In *Wearing Cultural Styles in Japan: Concepts of tradition and modernity in practice*, edited by C. S. Thompson and J. W. Traphagan. Albany, NY: SUNY Press.

Mol, A., I. Moser and J. Pols. 2010. 'Care: Putting practice into theory', *Care in Practice: On tinkering in clinics, homes and farms*, 8, 7–27. Bielefeld: Transcript Verlag.

Moloney, G., P. Holtz and W. Wagner. 2013. 'Editorial political cartoons in Australia: Social representations and the visual depiction of essentialism', *Integrative Psychological and Behavioral Science*, 47, 284–98.

Moore, G. and C. Atherton. 2020. 'Eternal forests: The veneration of old trees in Japan', *Arnoldia* 77(4): 24–31.

Moore, K. L. 2014. *The Joy of Noh: Embodied learning and discipline in urban Japan*. Albany, NY: SUNY Press.

Moore, K. L. 2017. 'A spirit of adventure in retirement: Japanese baby boomers and the ethos of interdependence', *Anthropology & Aging* 38(2): 10–28. https://doi.org/10.5195/aa.2017.159.

Moore, P. and A. Robinson. 2016. 'The quantified self: What counts in the neoliberal workplace', *New Media & Society* 18(11): 2774–92. https://doi.org/10.1177/1461444815604328.

Mort, M., T. Finch and C. May. 2009. 'Making and unmaking telepatients: Identity and governance in new health technologies', *Science, Technology, & Human Values* 34: 9–33.

Mueller, J. C. and J. McCollum. 2022. 'A sociological analysis of "OK Boomer"', *Critical Sociology* 48(2): 265–81. https://doi.org/10.1177/08969205211025724.

Murai, S. 2014. 'SoftBank unveils smartphone for seniors', *The Japan Times*. https://www.japantimes.co.jp/news/2014/10/28/business/corporate-business/softbank-unveils-smartphone-for-seniors/.

Murakami, K., R. Gilroy and J. Atterto. 2009. 'Planning for the ageing countryside in Japan: The potential impact of multi-habitation', *Planning Practice & Research* 24(3): 285–99. https://doi.org/10.1080/02697450903020734.

Murakami, Y., S. Kumon and S. Sato. 1979. *Bunmei Ttoshite no Ie-shakai (Familial Society as a Civilization)*. Tokyo: Chūōkōron.

Murthy, D. 2008. 'Digital ethnography: An examination of the use of new technologies for social research', *Sociology* 42(5): 837–55. https://doi.org/10.1177/0038038508094565.

Naito, T. and U. Gielen. 1992. '*Tatemae* and *honne*: A study of moral relativism in Japanese culture'. In *Psychology in International Perspective*, edited by Uwe P. Gielen, Leonore Lowe-Adler and Noach Milgram, 161–72 Boca Raton, FL: CRC Press.

Nakamura, M. 2020. 'Making an O-turn: The siren call of Japan's rural life', *The Japan Times*, 23 May 2020. https://www.japantimes.co.jp/life/2020/05/23/food/rural-life/.

Nakane, C. 1967. *Kinship and Economic Organization in Rural Japan*. London: Athlone Press.

Nakane, C. 1970. *Japanese Society*. Berkeley, CA: University of California Press.

Nakanishi, N. 1999. '"Ikigai" in older Japanese people', *Age and Ageing* 28(3): 323–4.

Nakano, L. 2004. *Community Volunteers in Japan: Everyday stories of social change*. Abingdon, Oxon; New York, NY: Routledge. https://doi.org/10.4324/9780203342299.

Nakatani, A. 2006. 'The emergence of "nurturing fathers": Discourses and practices of fatherhood in contemporary Japan'. In *The Changing Japanese Family*, edited by M. Rebick and A. Takenaka, 110–24. Abingdon, Oxon; New York, NY: Routledge.

Napier, S. J. 2018. *Miyazakiworld: A life in art*. New Haven, CN: Yale University Press.

National Institute of Population and Social Security Research. 2016. *Population Statistics*. http://www.ipss.go.jp/site-ad/index_english/Population%20%20Statistics.html. Accessed 23 February 2021.

National Institute of Population and Social Security Research. 2021. *Population Projections for Japan (2016–2065): Summary*. http://www.ipss.go.jp/pp-zenkoku/e/zenkoku_e2017/pp_zenkoku2017e_gaiyou.html#:~:text=According%20to%20the%20results%20of,and%20Figure%201%2D3. Accessed 23 February 2021.

Neff, G. and D. Nafus. 2016. *Self-tracking*. Cambridge, MA: MIT Press.

Negishi, M. 2014. 'Meet Shigetaka Kurita, the father of emoji', *Wall Street Journal*. https://www.wsj.com/articles/BL-JRTB-16473.

Neimeyer, R. A., D. Klass and M. R. Dennis. 2014. 'A social constructionist account of grief: Loss and the narration of meaning', *Death Studies* 38(8): 485–98.

Nelson, T. D. 2004. *Ageism: Stereotyping and prejudice against older persons*. Cambridge, MA: MIT Press.

Nemoto, K. 2013. 'Long working hours and the corporate gender divide in Japan', *Gender, Work & Organization* 20(5): 512–27.

NHK. 2019. '"Keimusho, Marude Kaigo Shisetsu ni. NHK Seiji Magajin"'. 21 August 2019. https://www.nhk.or.jp/politics/articles/feature/21325.html. Accessed 5 December 2023.

Nomura, S., H. Sakamoto, S. Glenn, Y. Tsugawa, S. K. Abe, M. M. Rahman, J. C. Brown, S. Ezoe, C. Fitzmaurice, T. Inokuchi and N. J. Kassebaum. 2017. 'Population health and regional variations of disease burden in Japan, 1990–2015: A systematic subnational analysis for the Global Burden of Disease Study 2015', *The Lancet* 390(10101): 1521–38.

Norbeck, E. 1977. 'Changing associations in a recently industrialized Japanese community', *Urban Anthropology* 6(1): 45–64.

Nozawa, S. 2015. 'Phatic traces: Sociality in contemporary Japan', *Anthropological Quarterly* 88(2): 373–400.

Obikwelu, F. E., K. Ikegami and T. Tsuruta. 2017. 'Factors of urban-rural migration and socio-economic condition of i-turn migrants in rural Japan', *Journal of Asian Rural Studies* 1(1), 70–80. https://doi.org/10.20956/jars.v1i1.727.

OECD. 2018. *Working Better with Age: Japan*. Organisation for Economic Co-operation and Development. https://www.oecd-ilibrary.org/social-issues-migration-health/working-better-with-age-japan_9789264201996-en.

OECD. 2019a. '"Health check-ups in Japan"'. In *OECD Reviews of Public Health: Japan: A healthier tomorrow*. Paris: OECD Publishing. https://doi.org/10.1787/9789264311602-7-en. Accessed 16 December 2023.

OECD. 2019b. *Pensions at a Glance 2019: OECD and G20 Indicators*. Organisation for Economic Co-operation and Development. https://www.oecd-ilibrary.org/social-issues-migration-health/pensions-at-a-glance-2019_b6d3dcfc-en.

OECD. 2020a. *Employment/Population Ratio by Sex and Age Group*. OECD Gender Equality. https://www.oecd.org/gender/data/employment/.

OECD. 2020b. *Key Charts on Employment*. OECD Gender Equality. https://www.oecd.org/gender/data/employment/.

OECD and World Health Organization. 2020c. *Health at a Glance: Asia/Pacific 2020: Measuring progress towards universal health coverage*. https://doi.org/10.1787/26b007cd-en.

OECD. 2024. *Elderly population (indicator)*. https://doi.org/10.1787/8d805ea1-en. Accessed 5 June 2024.

Ogasawara, Y., A. Kuroiwa, Y. Kondo, T. Kishino, T. Nakayama, K. Miyazi, Y. Fukutomi, F. Kawamura, N. Maeda and S. Sone. 2013. 'Study of risk factors for lifestyle-related health problems in men in their 40s and 50s in Kochi Prefecture', *Journal of the Japanese Association of Rural Medicine*, 611–17.

Ogawa, A. 2015. *Lifelong Learning in Neoliberal Japan: Risk, community, and knowledge*. Albany, NY: SUNY Press.

Ogawa, N. and R. D. Retherford. 1993. 'Care of the elderly in Japan: Changing norms and expectations', *Journal of Marriage and the Family*, 585–97.

Ogawa, N. and R. D. Retherford. 1997. 'Shifting costs of caring for the elderly back to families in Japan: Will it work?', *Population and Development Review*, 59–94.

Ohara, K. 2021. 'Wakayama: Best spot to let your "Ya-ho" shouts be heard', *Yumiuri Shimbun*. https://japannews.yomiuri.co.jp/features/japan-focus/20210626-51392/. Accessed 6 December 2023.

Ohashi, K., F. Kato and L. Hjorth. 2017. 'Digital genealogies: Understanding social mobile media LINE in the role of Japanese families', *Social Media + Society* 3(2). https://doi.org/10.1177/2056305117703815.

Ohno, A. 2005. *Sanson kankyo shakaigaku jyosetsu* ['Introduction to environmental sociology of mountain villages']. Tokyo: Nouson gyoson bunka kyokai.

Ohnuki-Tierney, E. and O.-T. Emiko. 1984. *Illness and Culture in Contemporary Japan: An anthropological view*. Cambridge: Cambridge University Press.

Okabe, Y., S. Shiratori and K. Suda. 2019. 'What motivates Japan's international volunteers? Categorizing Japan's Overseas Cooperation Volunteers (JOCVs)', *VOLUNTAS: International Journal of Voluntary and Nonprofit Organizations* 30(5): 1069–89. https://doi.org/10.1007/s11266-019-00110-x.

Okada, T. 2005. 'Youth culture and the shaping of Japanese mobile media: Personalization and the *keitai* internet as multimedia'. In *Personal, Portable, Pedestrian: Mobile phones in Japanese life*, edited by M. Ito, D. Okabe and M. Matsuda, 41–60. Cambridge, MA: MIT Press.

Okochi, K. 1989. 'A memorandum on the ground plan (May 1982) of the Tokyo municipal foundation for the promotion of undertakings for the aged'. March 1989. In *Living in an Aging Society: A digest of speeches by Kazuo Okochi*, edited by the National Silver Human Resource Centers Association, 9–10. Tokyo: National Silver Human Resource Centers Association.

Onitsuka, K. and S. Hoshino. 2018. 'Inter-community networks of rural leaders and key people: Case study on a rural revitalization program in Kyoto Prefecture, Japan', *Journal of Rural Studies* 61, 123–36. https://doi.org/10.1016/j.jrurstud.2018.04.008.

Ono, M. 2008. 'Long-stay tourism and international retirement migration: Japanese retirees in Malaysia', *Senri Ethnological Reports* 77, 151–62.

Ono, M. 2015. 'Commoditization of lifestyle migration: Japanese retirees in Malaysia', *Mobilities* 10(4): 609–27. https://doi.org/10.1080/17450101.2014.913868.

Oshio, T., E. Usui and S. Shimizutani. 2018. 'Labor force participation of the elderly in Japan', National Bureau of Economic Research. No. w24614. https://doi.org/10.3386/w24614.

Ouchi, Y., H. Rakugi, H. Arai, M. Akishita, H. Ito, K. Toba and I. Kai. 2017. 'Redefining the elderly as aged 75 years and older: Proposal from the Joint Committee of Japan Gerontological Society and the Japan Geriatrics Society'. (n.d.), *Geriatrics and Gerontology International* 17: 1045–7. https://doi.org/10.1111/ggi.13118.

Ozaki, M. 2014. 'Kochi's challenge: A prefecture tackling depopulation (part 2)'. In *JFS Newsletter* No.145, September 2014, edited by Junko Edahiro. https://www.japanfs.org/en/news/archives/news_id035061.html. Accessed 31 December 2023.

Ozawa-de Silva, C. 2018. 'Stand by me: The fear of solitary death and the need for social bonds in contemporary Japan'. In *The Routledge Handbook of Death and the Afterlife,* edited by C. K. Cann, 85–95. Abingdon, Oxon; New York, NY: Routledge.

Pandian, A. 2019. *A Possible Anthropology: Methods for uneasy times*. Durham, NC: Duke University Press.

Pekkanen, R. J., Y. Tsujinaka and H. Yamamoto. 2014. *Neighborhood Associations and Local Governance in Japan*. Abingdon, Oxon; New York, NY: Routledge. https://doi.org/10.4324/9781315797731.

Pelto, G. 2008. 'Taking care of children: Applying anthropology in maternal and child nutrition and health', *Human Organization*, 67(3), 237–43.

Pickard S. 2019. 'Old age and the neoliberal life course'. In *Interrogating the Neoliberal Lifecycle*, edited by B. Clack and M. Paule, 215–34. Cham: Palgrave Macmillan.

Pina-Cabral, J. and D. Theodossopoulos. 2022. 'Thinking about generations, conjuncturally: A toolkit', *The Sociological Review* 70(3): 455–73. https://doi.org/10.1177/0038026121 1062301.

Pink, S. 2001. 'More visualising, more methodologies: On video, reflexivity and qualitative research', *Sociological Review* 49(4): 586–99.

Pink, S (ed.). 2009. *Visual Interventions: Applied visual anthropology*. Oxford; New York: Berghahn.

Pink, S. 2011. 'Digital visual anthropology: Potentials and challenges'. In *Made to Be Seen: Perspectives on the history of visual anthropology*, edited by M. Banks and J. Ruby, 209–33. Chicago, IL: University of Chicago Press.

Pink, S. 2012. *Advances in Visual Methodology*. Thousand Oaks, CA: Sage. https://doi.org/10.4135/9781446250921.

Pink, S. 2015. *Doing Sensory Ethnography*. Thousand Oaks, CA: Sage.

Pink, S. and V. Fors. 2017. 'Self-tracking and mobile media: New digital materialities', *Mobile Media & Communication* 5(3): 219–38. https://doi.org/10.1177/2050157917695578.

Pink, S. and L. Hjorth. 2012. 'Emplaced cartographies: Reconceptualising camera phone practices in an age of locative media', *Media International Australia Incorporating Culture & Policy*, 145(1): 145–55. https://doi.org/10.1177/1329878X1214500116.

Plath, D. W. 1975. 'The last Confucian sandwich: Becoming middle aged', *Journal of Asian and African Studies* 10(1): 51.

Plath, D. W. 1980. *Long Engagements: Maturity in modern Japan* (Vol. 75). Redwood City, CA: Stanford University Press.

Plath, D. W. 1990. 'My-Car-isma: Motorizing the Showa Self', *Daedalus 119*(3): 229–44.

Plath, D. W. 2020. *The After Hours*. Berkeley, CA: University of California Press. https://doi.org/10.1525/9780520335288.

Pols, J. 2012. *Care at a Distance: On the closeness of technology*. Amsterdam: Amsterdam University Press.

Portacolone, E. 2011. 'Precariousness among older adults living alone in San Francisco: An ethnography'. Doctoral thesis. University of California, San Francisco.

Postill, J. and S. Pink. 2012. 'Social media ethnography: The digital researcher in a messy web', *Media International Australia* 145(1): 123–34. https://doi.org/10.1177/1329878X1214500114.

Qu, M., Coulton, T. and C. Funck. 2020. 'Gaps and limitations – Contrasting attitudes to newcomers and their role in a Japanese island community', *Bulletin of the Hiroshima University Museum* 12, 31–46. https://doi.org/10.15027/50631.

Quinn, W. and J. D. Turner. 2020. *Boom and Bust: A global history of financial bubbles*. Cambridge: Cambridge University Press.

Rattine-Flaherty, E. and A. Singhal. 2009. 'Analysing social-change practice in the Peruvian Amazon through a feminist reading of participatory communication research', *Development in Practice* 19(6): 726–36. https://doi.org/10.1080/09614520903026884.

Redfield, P. 2018. 'Intervention', 'Theorizing the contemporary', *Fieldsights*, 29 March, https://culanth.org/fieldsights/intervention.

Riach, K. 2007. '"Othering" older worker identity in recruitment', *Human Relations* 60(11): 1701–26. https://doi.org/10.1177/0018726707084305.

Rich, E. and A. Miah. 2014. 'Understanding digital health as public pedagogy: A critical framework', *Societies* 4: 296–315.

Rich, E. and A. Miah. 2017. 'Mobile, wearable and ingestible health technologies: Towards a critical research agenda', *Health Sociology Review* 26, 84–97.

Riordan, M. A. 2017. 'Emojis as tools for emotion work: Communicating affect in text messages', *Journal of Language and Social Psychology* 36(5): 549–67. https://doi.org/10.1177/0261927X17704238.

Roberson, J. E. 1995. 'Becoming *shakaijin*: Working-class reproduction in Japan', *Ethnology*, 34(4): 293–313. https://doi.org/10.2307/3773943.

Robertson, Jennifer. 1991. *Native and Newcomer: Making and remaking a Japanese City*. Berkeley. CA; Oxford: University of California Press.

Rolnik, S. 2011. 'The geopolitics of pimping'. In *Critique of Creativity: Precarity, subjectivity and resistance in the 'creative;ndustries'*, edited by U. Wuggenig, G. Raunig and G. Ray, 23–40. London: Mayfly. https://dspace-libros.metabiblioteca.com.co/display-item.jsp.

Roquet, P. 2016. *Ambient Media: Japanese atmospheres of self*. Minneapolis, MN: University of Minnesota Press. https://doi.org/10.5749/j.ctt19qggn7.

Rosenberger, N. 1994. 'Introduction'. In *Japanese Sense of Self* (Vol. 2), 1–20, edited by N. R. Rosenberger. Cambridge: Cambridge University Press.

Rosenberger, N. 2006. 'Young women making lives in northeast Japan'. In *Wearing Cultural Styles in Japan: Concepts of tradition and modernity in practice*, edited by C. S. Thompson and J. W. Traphagan, 76–95. Albany, NY: SUNY Press.

Rosenberger, N. 2017. 'Young organic farmers in Japan: Betting on lifestyle, locality, and livelihood', *Contemporary Japan* 29(1): 14–30.

Rowe, J. W. and R. L. Kahn. 1997. 'Successful aging', *The Gerontologist* 37(4): 433–40.

Rowland, D. T., 2012. *Population Aging: the transformation of societies* (Vol. 3). Heidelberg; New York, NY; London: Springer Science & Business Media.

Rubinstein, R. L. and K. de Medeiros. 2015. '"Successful aging", gerontological theory and neoliberalism: A qualitative critique', *The Gerontologist* 55(1): 34–42.

Ruby, J. (ed.). 2011. *Made to be Seen: Perspectives on the history of visual anthropology*. Chicago, IL: University of Chicago Press.

Ruckenstein, M. 2015. 'Uncovering everyday rhythms and patterns: Food tracking and new forms of visibility and temporality in health care', *Studies in Health Technology and Informatics* 215, 28–40.

Ruckenstein, M. and N. D. Schüll. 2017. 'The datafication of health', *Annual Review of Anthropology* 46, 261–78.

Russell, J. 2019. 'Chat app Line injects $182M into its mobile payment business', *TechCrunch*. https://techcrunch.com/2019/02/04/line-pay/?guccounter=1.

Saino, T. 1997. *Population Aging in Metropolitan Cities*. Tokyo: Taimeido.

Saito, J., M. Haseda, A. Amemiya, D. Takagi, K. Kondo and N. Kondo. 2019. 'Community-based care for healthy ageing: Lessons from Japan', *Bulletin of the World Health Organization* 97(8): 570–74. https://doi.org/10.2471/BLT.18.223057.

Sakaiya, T. 1976. *Dankai no sedai (Dankai Generation)*. Tokyo: Bungeisyunju.

Sakuno, H. 2006. 'The problems and expectations of regional development in hilly-mountainous region and correspondence of rural settlements', *Annals of the Japan Association of Economic Geographers* 52, 264–82.

Sanjek, R. and S. W. Tratner (eds). 2016. *eFieldnotes: The makings of anthropology in the digital world*. Philadelphia, PA: University of Pennsylvania Press.

Sasaki, L., L. Haapio-Kirk and Y. Kimura. 2021. 'Sharing virtual meals among the elderly: An ethnographic and quantitative study of the role of smartphones for distanced social eating in rural Japan'. In *Japanese Review of Cultural Anthropology*, vol. 21, 7–47.

Schwartz, F. 2003. 'Introduction: Recognizing civil society in Japan'. In *The State of Civil Society in Japan*, edited by F. J. Schwartz and S. J. Pharr, 1–20. Cambridge: Cambridge University Press. https://doi.org/10.1017/CBO9780511550195.002.

Schwarzenegger, C. 2020. 'Personal epistemologies of the media: Selective criticality, pragmatic trust, and competence–confidence in navigating media repertoires in the digital age', *New Media & Society* 22(2), 361–77. https://doi.org/10.1177/1461444819856919.

Schüll, N. D. 2016a. 'Data for life: Wearable technology and the design of self-care', *BioSocieties* 11:317–33.

Schüll N. D. 2016b. 'Tracking'. In *Experience: Culture, Cognition, and the Common Sense*, edited by C. A. Jones, D. Mather and R. Uchill, 195–203. Cambridge, MA: MIT Press.

Sekine, S. 2023. 'White Paper: Mutual support needed to combat rise in isolation', *The Asashi Shimbun*. https://www.asahi.com/ajw/articles/14983330.

Serafinelli, E. 2018. *Digital Life on Instagram: New social communication of photography*. Emerald Publishing Limited. https://doi.org/10.1108/9781787564954.

Shakuto, S. 2017. 'Ageing with bad-boy charm: An affective analysis of Japanese retirement migration in Malaysia', *Japanese Review of Cultural Anthropology* 18(1): 159–72.

Shakuto, S. 2018. 'An independent and mutually supportive retirement as a moral ideal in contemporary Japan', *The Australian Journal of Anthropology* 29(2): 184–94. https://doi.org/10.1111/taja.12277.

Shankar, A., A. McMunn, J. Banks and A. Steptoe. 2011. 'Loneliness, social isolation, and behavioral and biological health indicators in older adults', *Health Psychology* 30(4): 377–85. https://doi.org/10.1037/a0022826.

Sheller, M. and J. Urry. 2006. 'The new mobilities paradigm', *Environment and Planning A: Economy and space* 38(2): 207–26. https://doi.org/10.1068/a37268.

Shirai, K., H. Iso, H. Fukuda, Y. Toyoda, T. Takatorige and K. Tatara. 2006. 'Factors associated with "*ikigai*" among members of a public temporary employment agency for seniors (Silver Human Resources Centre) in Japan; gender differences', *Health and Quality of Life Outcomes* 4, 12.

Simpson, N. 2022. '*Kamzori*: Aging, care, and alienation in the post-pastoral Himalaya', *Medical Anthropology Quarterly* 36: 391–411. https://doi.org/10.1111/maq.12707.

Sinanan, J. 2019. 'Visualising intimacies: The circulation of digital images in the Trinidadian context', *Emotion, Space and Society* 31, 93–101. https://doi.org/10.1016/j.emospa.2019.04.003.

Sinanan J., L. Hjorth, K. Ohashi and F. Kato. 2018. 'Mobile media photography and intergenerational families', *International Journal of Communication* 12(17): 4106–22. https://ijoc.org/index.php/ijoc/article/viewFile/9666/2477.

Smith, C. 2022. '65 Amazing LINE Statistics (Chat App)', *DMR (Formerly Digital Marketing Ramblings)* [blog]. https://expandedramblings.com/index.php/line-statistics/.

Smith, R. J. 1978. *Kurusu: The price of progress in a Japanese village, 1951–1975*. Folkstone: Dawson.

Smith, R. J. and E. L. Wiswell. 1982. *The Women of Suye Mura*. Chicago. IL: University of Chicago Press. https://press.uchicago.edu/ucp/books/book/chicago/W/bo42710532.html.

Smith, S. G., S. E. Jackson, L. C. Kobayashi and A. Steptoe. 2018. 'Social isolation, health literacy, and mortality risk: Findings from the English Longitudinal Study of Ageing', *Health Psychology* 37(2): 160–69. https://doi.org/10.1037/hea0000541.

Sone, T., N. Nakaya, K. Ohmori, T. Shimazu, M. Higashiguchi, M. Kakizaki and I. Tsuji. 2008. 'Sense of life worth living (*ikigai*) and mortality in Japan: Ohsaki study', *Psychosomatic Medicine* 70(6): 709–15.

Stark, L. and K. Crawford. 2015. 'The conservatism of emoji: Work, affect, and communication', *Social Media + Society* 1(2). https://doi.org/10.1177/2056305115604853.

Statistics Bureau of Japan. 2015. *Reiwa 2nd year census*. https://www.stat.go.jp/data/kokusei/2020/index.html. Accessed 15 August 2020.

Statistics Bureau of Japan. 2021. Home page/Statistical Handbook of Japan 2021. https://www.stat.go.jp/english/data/handbook/c0117.html.

Statistics Bureau of Japan. 2022. *Employment Status Survey. 2022 Survey: Outline.* https://www.stat.go.jp/english/data/shugyou/2022/sum2022.pdf.

Steinberg, M. 2020. 'LINE as Super App: Platformization in East Asia', *Social Media + Society* 6(2). https://doi.org/10.1177/2056305120933285.

Steptoe, A., A. Deaton and A. A. Stone. 2015. 'Subjective wellbeing, health, and ageing', *The Lancet* 385(9968): 640–8.

Stevenson, L. 2014. *Life Beside Itself: Imagining care in the Canadian Arctic* (1st ed.). Berkeley, CA: University of California Press.

Stewart, K. 2007. *Ordinary Affects*. Durham, NC: Duke University Press. https://doi.org/10.1515/9780822390404.

Stoller, P. 1997. *Sensuous Scholarship*. Philadelphia, PA: University of Pennsylvania Press.

Stoller, P. 2010. *The Taste of Ethnographic Things: The senses in anthropology*. Philadelphia, PA: University of Pennsylvania Press. https://doi.org/10.9783/9780812203141.

Suehara, T. 2006. 'Labor exchange systems in Japan and DR Congo: Similarities and differences', *Journal of Comparative Social Welfare* 9(1): 11.

Sugimoto, Y. 2020. *An Introduction to Japanese Society*. Cambridge: Cambridge University Press.

Suzuki, N. 2012. 'Creating a community of resilience: New meanings of technologies for greater well-being in a depopulated town', *Anthropology & Aging* 33(3): 87–96.

Suzuki, N. 2018. 'Creating an age-friendly community in a depopulated town in Japan'. In *The Global Age-Friendly Community Movement: A critical appraisal*, edited by P. B. Stafford, 229–46. Oxford and New York: Berghahn Books.

Suzuki, N. 2020. 'Weaving flexible aging-friendly communities across generations while living with COVID-19', *Anthropology and Aging* 41(2): 155–66. http://dx.doi.org/10.5195/aa.2020.311.

Suzuki, T. 2015. 'After neoliberalism?', *Asian Journal of Social Science* 43(1–2): 151–77. https://doi.org/10.1163/15685314-04301008.

Swane, C. E. 2017. 'Agency through media in the everyday life of nursing home residents', *Innovation in Aging* 1(Suppl. 1), 199.

Swift, P. 2017. 'Cosmetic cosmologies in Japan: Notes towards a superficial investigation'. In *Cosmopolitics: The collected papers of the Open Anthropology Cooperative*, volume I, 140, edited by J. Shaffner and H. Wardle. St Andrews: Open Anthropology Cooperative Press.

Tacchi, J., K. R. Kitner and K. Crawford. 2012. 'Meaningful mobility: Gender, development and mobile phones', *Feminist Media Studies* 12(4): 528–37. https://doi.org/10.1080/14680777.2012.741869.

Takahashi, E. and J. Danely. 2020. 'Those who come early: Reflections on the social standing of senior citizens in the time of the COVID-19 pandemic in Japan', *Somatosphere*, 31 May 2020. https://somatosphere.com/2020/those-who-come-early.html/.

Takami, T. 2018. 'Gender segregation at work in Japanese companies: Focusing on gender disparities in desire for promotion', *Japan Labor Issues* 2(11): 7–12.

Takaragawa, S., T. Smith, K. Hennessy, P. Astacio, J. Chio, C. Nye and S. Shankar. 2019. 'Bad habitus: Anthropology in the age of the multimodal', *Sociological Anthropology*. https://doi.org/10.1111/aman.13265.

Takeda, S. 2020. 'Fluidity in rural Japan: How lifestyle migration and social movements contribute to the preservation of traditional ways of life on Iwaishima'. In *Japan's New Ruralities: Coping with decline in the periphery*, edited by W. Manzenreiter, R. Lützeler and S. Polak-Rottmann, 196–211. Abingdon, Oxon; New York, NY: Routledge.

Tanaka, K. and N. Johnson. 2021. *Successful Aging in a Rural Community in Japan*. Durham, NC: Carolina Academic Press.

Taniguchi, H. 2010. 'Who are volunteers in Japan?', *Nonprofit and Voluntary Sector Quarterly* 39(1): 161–79. https://doi.org/10.1177/0899764008326480.

Tanno, K., K. Sakata, M. Ohsawa, T. Onoda, K. Itai, Y. Yaegashi, A. Tamakoshi and JACC Study Group. 2009. 'Associations of *ikigai* as a positive psychological factor with all-cause mortality and cause-specific mortality among middle-aged and elderly Japanese people: Findings from the Japan Collaborative Cohort Study', *Journal of Psychosomatic Research* 67(1): 67–75.

Tateno, M., A. R. Teo, W. Ukai, J. Kanazawa, R. Katsuki, H. Kubo and T. A. Kato. 2019. 'Internet addiction, smartphone addiction, and *hikikomori* trait in Japanese young adults: Social isolation and social network', *Frontiers in Psychiatry* 10. https://www.frontiersin.org/article/10.3389/fpsyt.2019.00455.

Tatsuki, S. 2000. 'The Kobe earthquake and the renaissance of volunteerism in Japan', *Journal of Kwansei Gakuin University Department of Sociology Studies* 87, 185–96.

Taussig, M. 2011. *I Swear I Saw This: Drawings in fieldwork notebooks, namely my own*. Chicago, IL: University of Chicago Press.

Taylor, M. and I. Taylor. 1995. *Writing and Literacy in Chinese, Korean and Japanese*. Studies in Written Language and Literacy 3. Amsterdam: John Benjamins Publishing Company. https://benjamins.com/catalog/swll.3.

Taylor, T. L., T. Boellstorff, B. Nardi and C. Pearce. 2012. *Ethnography and Virtual Worlds: A handbook of method*. Princeton, NJ: Princeton University Press.

The Care Collective: A. Chatzidakis, J. Hakim, J. Litter, C. Rottenberg and L. Segal. 2020. *The Care Manifesto*. London; Brooklyn, NY: Verso.

The Economist. 2019. 'In Japan, there is a boom in books by and for the elderly', *The Economist*. https://www.economist.com/asia/2019/02/23/in-japan-there-is-a-boom-in-books-by-and-for-the-elderly.

The Government of Japan. 2005. English translation of the 'Law concerning Nippon Telegraph and Telephone Corporation, etc' [English translation]. https://www.soumu.go.jp/main_sosiki/joho_tsusin/eng/Resources/laws/NTTLaw.pdf.

The Government of Japan. 2022. 'Yui', *Kizuna*. Tomodachi Disaster Prevention Edition 2015. https://www.japan.go.jp/tomodachi/2015/disaster_prevention_edition_2015/yui.html.

Theodossopoulos, D. 2024. 'Graphic ethnography'. In *The International Encyclopedia of Anthropology*, edited by H. Callan. https://doi.org/10.1002/9781118924396.wbiea2474.

Thomas, J. M. and J. G. Correa. 2015. *Affective Labour: (Dis) assembling distance and difference*. Lanham, MD: Rowman & Littlefield International.

Thompson, J. B. 2005. 'The new visibility', *Theory, Culture & Society* 22(6): 31–51.

Tiefenbach, T. and F. Kohlbacher. 2013. 'Happiness and life satisfaction in Japan by gender and age', Tokyo: German Institute for Japanese Studies.

Tosa Town Office. 2020. *Kōchi Prefecture Tosa Town, official home page*. http://www.town.tosa.kochi.jp/.

Toyota, M. 2006. 'Ageing and transnational householding: Japanese retirees in Southeast Asia', *International Development Planning Review* 28(4): 515–31.

Traphagan, J. W. 1998. 'Contesting the transition to old age in Japan', *Ethnology* 37(4): 333–50. https://doi.org/10.2307/3773786.

Traphagan, J. W. 2004. *The Practice of Concern: Ritual, well-being, and aging in rural Japan*. Durham, NC: Carolina Academic Press.

Traphagan, J. W. 2006. 'Power, family, and filial responsibility related to elder care in rural Japan', *Care Management Journals* 7(4): 205–12.

Traphagan, J. W. 2020. *Cosmopolitan Rurality, Depopulation, and Entrepreneurial Ecosystems in 21st-Century Japan*. Amherst, NY: Cambria Press.

Traphagan, J. W. and J. Knight. 2003. *Demographic Change and the Family in Japan's Aging Society*. Albany, NY: SUNY Press.

Tsugane, S. 2021. 'Why has Japan become the world's most long-lived country: Insights from a food and nutrition perspective', *European Journal of Clinical Nutrition* 75(6): 921–8.

Tsutsui, T. and N. Muramatsu. 2005. 'Care-needs certification in the long-term care insurance system of Japan', *Journal of the American Geriatrics Society* 53(3): 522–7.

Tsutsui, T., N. Muramatsu and S. Higashino. 2014. 'Changes in perceived filial obligation norms among coresident family caregivers in Japan', *The Gerontologist* 54(5): 797–807.

Tu, W. and W. Du. 1996. *Confucian Traditions in East Asian Modernity: Moral education and economic culture in Japan and the four mini-dragons*. Cambridge, MA: Harvard University Press.

Turkle, S. 2011. *Alone Together: Why we expect more from technology and less from each other*. New York, NY: Basic Books.

Turkle, S. 2012. 'Opinion | The flight from conversation'. *The New York Times*. 22.04.2012. https://www.nytimes.com/2012/04/22/opinion/sunday/the-flight-from-conversation.html.

Turner, E. 2012. *Communitas: The anthropology of collective joy*. Basingstoke; New York, NY: Palgrave Macmillan.

Turner, V. 1969. *The Ritual Process: Structure and anti-structure*. Chicago, IL: Aldine Publishing

Twenge, J. M., J. Haidt, A. B. Blake, C. McAllister, H. Lemon and A. Le Roy. 2021. 'Worldwide increases in adolescent loneliness', *Journal of Adolescence* 93, 257–69. https://doi.org/10.1016/j.adolescence.2021.06.006.

Ullrich, M. 2017. 'Media use during escape. A contribution to refugees' collective agency'. https://doi.org/10.25969/MEDIAREP/3854.

United Nations. 2019. 'Ageing'. https://www.un.org/en/global-issues/ageing. Accessed 22 March 2022.

United Nations. 2020. Department of Economic and Social Affairs & Population Division. *World Population Ageing 2020 Highlights: Living arrangements of older persons*. https://www.un.org/development/desa/pd/sites/www.un.org.development.desa.pd/files/undesa_pd-2020_world_population_ageing_highlights.pdf. Accessed 22 March 2022.

Valtorta, N. K., M. Kanaan, S. Gilbody, S. Ronzi and B. Hanratty. 2016. 'Loneliness and social isolation as risk factors for coronary heart disease and stroke: Systematic review and meta-analysis of longitudinal observational studies', *Heart*, 102(13): 1009–16. https://doi.org/10.1136/heartjnl-2015-308790.

Van der Horst, M. and S. Vickerstaff. 2022. 'Is part of ageism actually ableism?', *Ageing and Society* 42(9): 1979–1990. https://doi.org/10.1017/S0144686X20001890.

Van Dijck, J. 2014. 'Datafication, dataism and dataveillance: Big Data between scientific paradigm and ideology', *Surveillance & Society* 12(2): 197–208.

Van Dyk, S. 2016. 'The othering of old age: Insights from postcolonial studies', *Journal of Aging Studies* 39, 109–20. https://doi.org/10.1016/j.jaging.2016.06.005.

Van Gennep, A. 2019 [1960]. *The Rites of Passage* (2nd ed.). Translated by Monika B. Vizedom and Gabrielle L. Caffee. Chicago, IL: The University of Chicago Press.

Verbruggen, C., B. M. Howell and K. Simmons. 2020. 'How we talk about aging during a global pandemic matters: On ageist othering and aging "others" talking back', *Anthropology & Aging* 41(2), 230–45.

Villi, M. 2012. 'Visual chitchat: The use of camera phones in visual interpersonal communication', *Interactions: Studies in Communication & Culture* 3(1): 39–54. https://doi.org/10.1386/iscc.3.1.39_1.

Vogel, E. F. 2020 [1963]. *Japan's New Middle Class: The salary man and his family in a Tokyo suburb*. Berkeley, CA: University of California Press.

Wada, S. 2000. 'Kōreishakai Ni Okeru "Ikigai" No Ronri' [The logic of 'ikigai' in aging society], *Ikigai Kenkyū [Study of Ikigai]* 12: 18–45.

Wakita, H. and D. P. Phillips. 1993. 'Women and the creation of the "Ie" in Japan: An overview from the medieval period to the present', *U.S.-Japan Women's Journal. English Supplement* 4, 83–105.

Walton, S. 2015. 'Re-envisioning Iran online: Photoblogs and the ethnographic "digital-visual moment"', *Middle East Journal of Culture and Communication* 8(2–3): 398–418. https://doi.org/10.1163/18739865-00802012.

Walton, S. 2021. *Ageing with Smartphones in Urban Italy: Care and community in Milan and beyond*. London: UCL Press. https://doi.org/10.14324/111.9781787359710.

Walton, S. and L. Haapio-Kirk. 2021. 'Doing multimodal anthropology of ageing with smartphones. Cases from Italy and Japan', *Anthrovision. Vaneasa Online Journal* (9.2).

Wang, X. 2016. *Social Media in Industrial China*. London: UCL Press. https://doi.org/10.14324/111.9781910634646.

Warburton, T. 2005. 'Cartoons and teachers: Mediated visual images as data'. In *Image-based Research*, edited by J. Prosser, 252–62. Abingdon, Oxon; New York, NY: Routledge.

Watanabe, H. R. 2020. 'Low levels of digitalisation are a barrier to telework in Japan', *East Asia Forum*. 20 June 2020. https://www.eastasiaforum.org/2020/06/20/low-levels-of-digitalisation-are-a-barrier-to-telework-in-japan/. Accessed 16 December 2023.

Watson, A. and D. Lupton. 2022. 'Remote fieldwork in homes during the COVID-19 pandemic: Video-call ethnography and map drawing methods', *International Journal of Qualitative Methods* 21. https://doi.org/10.1177/16094069221078376.

Weicht, B. 2013. 'The making of "the elderly": Constructing the subject of care', *Journal of Aging Studies* 27(2): 188–97. https://doi.org/10.1016/j.jaging.2013.03.001.

Weiss, R. S., S. A. Bass, H. K. Heimovitz and M. Oka. 2005. 'Japan's silver human resource centers and participant well-being', *Journal of Cross-Cultural Gerontology* 20(1): 47–66.

White, L. 2018. *Gender and the Koseki in Contemporary Japan: Surname, power, and privilege*. Abingdon, Oxon; New York, NY: Routledge.

White, M. 1992. 'Home truths: Women and social change in Japan', *Daedalus* 121(4): 61–82.

WHO. 2011. *mHealth – New Horizons for Health through Mobile Technologies*. World Health Organization. https://www.afro.who.int/publications/mhealth-new-horizons-health-through-mobile-technologies. Accessed 2 January 2022.

Wilson, R. S., K. R. Krueger, S. E. Arnold, J. A. Schneider, J. F. Kelly, L. L. Barnes and D. A. Bennett. 2007. 'Loneliness and risk of Alzheimer's disease', *Archives of General Psychiatry* 64(2): 234.

Wittfogel, K. A. 1957. *Oriental Despotism: A comparative study of total power*. Oxford: Oxford University Press.

Yamakubi, N. 2019. Head of the Tosa-chō Social Welfare Council (*Shakai-Fukushi-Kyogikai*). Personal communication, 1 February 2019.

Yanagita, K. 1967. *Meiji Taishō Shi: Sesō Hen* (History of the Meiji and Taishō Eras: The phases of people's life). In Japanese. Tokyo: Tōyō bunko.

Yoshikawa, H. 2021. *Ashes to Awesome: Japan's 6,000-Day economic miracle*. Tokyo: Japan Publishing Industry Foundation for Culture.

Zhou, S. 2020. 'Evaluating the impact of labor market and pension policies on the rise of elderly female poverty in Japan', *Social Impact Research Experience (SIRE)*. https://repository.upenn.edu/sire/78.

Index

generational attitudes towards 223–5
graphic methodology for the study
of 227–8
negative attitudes towards 216, 225
information 118, 123, 135, 148
health 167, 183
personal 207
'informational capital' 252
intergenerational
care 19, 54, 191
co-residence 49, 56, 64
interdependence 40
relations 11, 16, 53, 191, 193, 209, 245
tensions 40
internet 14–15, 122, 131–4, 136, 139
installation of high speed 189
negative implications of 105–6
point of entry to 154, 257
interpersonal communication 137, 147–9
intimacy
digital 57, 67, 121, 206
mediation of 39, 53, 156
invisibility, tactical 17, 114, 124, 159, 161,
241, 246
isolation
at home 39, 215
digital response to 181–5
due to lack of digital access 54, 126
due to precarious working conditions 89
due to population decline and the
pandemic 100
era of 105, 201
risk of 65, 113, 168, 194, 257

Japan
ageing society 2, 7, 11, 40–2
post-war economic rise of 2, 8–10
recession in 10, 75, 167, 216, 243
technology development 133–7
journey, life 227–9, 231
judgement 44, 46–7, 65, 122, 149

kaomoji 139
kawaii (cute) 2, 145
kinship 101, 103, 114
digital 245
fictive 196
knowledge
devaluation of 40
passing on 195, 199, 228
production 31–2
self- 168, 173, 177
Kōchi
culture of pilgrimage 179
dialect 36n104
fieldsite 18, 23
map of 24f1.5
prefectural government 170, 172, 178–9
short video introducing fieldsite 23f1.4
Kyoto
festivals 23
fieldsite 18–22
map of 22f1.3
neighbourhood sociality 106–12
people 123
short video introducing fieldsite 20f1.2

labour (*see also* non-retirement)
affective 57, 89, 132, 160, 252
domestic 94–5, 220, 225
force among older adults 76, 95–6
gendered 50, 67, 69n58, 73–5, 77–81, 243
precarious (*see also* precarity) 89, 216
shortage 76, 83
life course 43–4, 61, 63, 65, 213–14
life purpose (*see ikigai*)
life story 228
LINE (*see* applications)
living well 18, 114, 232, 258
loneliness (*see also* isolation) 89, 100, 126,
151, 256
reducing 171, 182
longevity 12, 27, 40, 43, 46, 66, 214, 215
Long Term Care Insurance (LTCI) 50,
167, 171
loss 20, 229
of agency 42, 199
of community 110
of connection 171

mHealth
critique of 167, 185
definitions of 165, 167
ethnographic example of 165–6
government example of 178–81
smartphone food club, intervention 181–5
marriage 44, 49, 63, 69n58, 77
methodology
graphic 30–2, 151–4, 162n26, 248–54
participant observation 28
short video describing research methods
28f1.6
middle-aged 43, 50–1, 103, 233
migration (*see also* Regional Revitalisation
Cooperation Officer Program)
domestic migrant (*ijū-sha*) 5, 34n9
'I-turner' 5, 34n9, 191–2, 201, 202, 205,
207–8
lifestyle 191, 203–5
motivations for 5, 9, 11, 26, 99, 189–91
physical and digital 205–9
'U-turner' 5, 26, 34n9, 196, 207, 221, 246
misinformation 119
modernity 26, 103, 194
morality
around ageing 75, 77, 84, 95, 100,
216, 255
judgements 46
moral citizen 66
of care 79, 80, 84
motivation 175–6, 204
muen shakai (relationless society) 105,
125, 245
multigenerational households 49–51
digital example of 56–60
ethnographic example of 51–56
multimodal research methods
(*see* methodology)
mutual aid (*see also* reciprocity) 101–2, 171

narratives
graphic 30–1, 250–1
of ageing 12, 42, 213–14

urbanisation (*see also* depopulation) 8, 191
urgency 101, 195, 199, 256
user-friendly 158

video
  short video introducing Kyoto fieldsite 20f1.2
  short video introducing rural fieldsite in
    Kōchi Prefecture 23f1.4
  short video describing fieldwork methods
    28f1.6
  short video about Miyagawa-san's hobby of
    building model ships 117f4.2
  short video about Hayashi-san's
    quilt-making 121f4.4
  short video about smartphones as a lifeline
    136f5.2
  short video about the annual health check
    in Tosa-chō 174,f6.1

short video about *ikigai* as finding work to
  dedicate one's life to 218f8.1
visibility (*see also* invisibility) 17, 124, 144,
  189–90, 207, 231
visual culture 18, 253
visual digital communication 14–18, 132, 137,
  140, 144–6, 242
volunteer (*borantia*) 86–7, 100–3, 215
voice-to-text feature 136
vulnerability 42, 133, 136, 256

welfare provision 45, 170
wellbeing 26, 40, 103, 172–3, 193, 214,
  215–19
women, role and status of 9, 49, 50, 59–60,
  75. 77–81, 88
work *see* employment

* 9 7 8 1 7 8 7 3 5 5 7 7 4 *